Math Light

Basic and Intermediate Algebra

Dilshad Akrayee

This book may help you pass the following tests:

First
Edition

1) GED
2) SAT
3) GRE
4) ACT
5) College Algebra

Acknowledgment and Dedication

This book is dedicated to my loving wife, Hilbeen, and my two precious kids, Heleen and Hari, for having the patience with me and taking a challenge that decreases the amount of time I can spend with them. The three of you are the light and love of my life. Special thanks go to Widad Akrayee, my mother and best teacher, who took a lot of time to edit and revise my book. Thank you mother for all your support!

In memory of my uncle, Raoof Kamil Raoof, who left the first fingerprints of writing on my life. Uncle, you will never be forgotten.

About the Author

The author of this book, Dilshad Akrayee, believes in dreams that gradually turn into goals and then goals be achieved. So, since he was a student at middle school, writing a book whether a math book or any others was one of the priorities that he ever dreamed of. He knows and encourages that anyone who has a dream should have a goal to be achieved sooner or later. Dilshad Akrayee holds a master degree from Shorter University and degrees in mathematics from Mosul University (B.S.), and a diploma in Draft Engineering from North Western Technical College in Rome, GA in addition to several other certificates such as AutoCAD from Draft Tech Systems Inc. in Atlanta, GA.

Dilshad has been an Instructor of Developmental Mathematics at Georgia Highlands College since 2007 and he was also a Math Tutor at Georgia Northwestern College. Teaching math and physics has been Dilshad's inspiring and considerate career in Kurdistan, Guam Island, and the United States for more than ten years. His dedication and commitment have been critical to the success of his students scientifically and academically. His extra effort has foster self and administration that provided his students' math skills that are essential to their practical life and future.

Since 2007 he started, organized, researched and planned to write this book. The book should help students to pass college algebra, GRE, GED, ACT and SAT tests.

Dilshad Akrayee, MBA/B. S, Mathematics

Math Instructor at Georgia Highlands College. Rome, GA

Former Math Teacher at Georgia Northwestern Technical College/Adult Education. Rome, Georgia

http://www.highlands.edu/site/faculty-dilshad-akrayee

Comments or Suggestions

If you have any comments or suggestions about this book, please e-mail the author at dakrayee@highlands.edu

Table of Contents

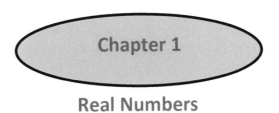

Chapter 1

Real Numbers

Section 1.1 Addition, Subtraction, Multiplication, and Division of Real Numbers

Basic things you could do with the calculator

+	Add signed numbers
-	Subtract signed numbers
x	Multiply signed numbers
÷	Divide signed numbers
√	Find the square root of a number
∧	Find the value of a number raised to a rational exponent. (i.e. $9^{3/2} = 27$)

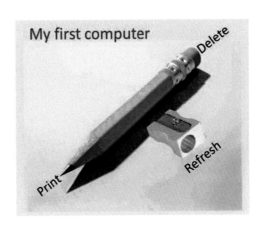

My first computer

Delete

Refresh

Print

Addition

Adding two (or more) numbers means to find their sum (or total). The symbol used for addition is '+'.

➢ **Adding two numbers with the same sign:**

1. **Put the same sign.**

2. **Add these two numbers.**

Example 1) Add

(+ 8) + (+ 6) = + 14

Hint: 8 and 6 have the same sign, which is **+**.

Example 2) Add

(+7) + (+4) = + 11

Hint: 7 and 4 have the same sign, which is **+**.

Example 3) Add

$$(-9) + (-6) = -15$$

Hint: 9 and 6 have the same sign, which is **−**.

Example 4) Add

$$(-8) + (-2) = -10$$

Hint: 8 and 2 have the same sign, which is **−**.

➢ **Adding two numbers with the different signs:**

1. Put the sign of the biggest number.
2. Big number – Small number.

Example 5) Add

$$(+8) + (-6) = +2$$

Here we put **+** because **+** is the sign of big # which is 8.

(Big number) – (Small number) = 8 – 6 = **2**.

Example 6) Add

$$(+8) + (-9) = -1$$

Here we put **−** because **−** is the sign of big # which is 9.

(Big number) – (Small number) = 9 – 8 = 1.

Example 7) Add

$$(+8) + (-19) = -11$$

Here we put **−** because **−** is the sign of big # which is 19.

(Big number) – (Small number) = 19 – 8 = **11**.

Example 8) Add

(+ 28) + (−26) = + 2

Here we put + because + is the sign of big # which is 28.

(Big number) – (Small number) = 28 – 26 = 2.

Subtraction

Subtracting one number from another is to find the difference between them. The symbol used for subtraction is '−'. This is known as the minus sign.

For example, 17 – 8 = 9

This is read as seventeen take away eight is equal to nine or we can say that 17 minus 8 is 9.

Example 9) Subtract 9 from 16.

<u>Solution:</u>

16 – 9 = 7 usually we put the number (which is 16 in this example) after the word "from" at beginning.

Multiplication

Multiplication means times (or repeated addition). The symbol used for multiplication is '×'.

For example, 7 × 2 = 14
This is read as seven times two is equal to fourteen or simply, seven times two is fourteen.

To multiply a large number with another number, we write the numbers vertically and generally multiply the larger number with the smaller number.

➢ **Multiplying two numbers with the same signs**

　　1. Put +.

　　2. Multiply the numbers.

Example 10) Multiply

(+ 8) * (+ 6) = + 48

Here we put + sign because both 8 and 6 have the same

sign which is + then 8 times 6 is 48.

Example 11) Multiply

(- 7) * (- 6) = + 42

Here we put + sign because both 7 and 6 have the same sign which is – then 7 times 6 is 42.

> **Multiplying two numbers with different signs**

3. Put –.

4. **Multiply the numbers.**

Example 12) Multiply

(+ 8) * (− 7) = − 56

You can also consider + as good and – as bad

(+) (+) = (+) when something good happens to somebody good... that's good.

(+) (-) = (-) when something good happens to somebody bad ...that's bad.

(-) (+) = (-) when something bad happens to somebody good ...that's bad.

(-) (-) = (+) when something bad happens to somebody bad ...that's good.

Division

Division 'undoes' multiplication and involves a number called the dividend being 'divided' by another number called the divisor. The symbols used for division are '÷' or ' / '

It is obivious that $9 * 5 = 45$ therefore, $45 \div 5 = 9$.

Note: Long division is an algorithm that repeats the basic steps of

1) Divide   **2)** Multiply   **3)** Subtract   **4)** Drop down the next digit   **5)** Start over

To remember these steps, you memorize the statement Dirty Monkeys Smell Bad Sometimes (DMSBS)

Where;

D for Division

M for Multiplication

S for Subtraction

B for Bring it down

S for Start over

Example 13) Calculate $692 \div 4$

<u>Solution:</u>

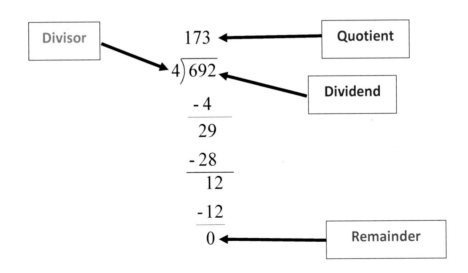

> **Dividing two numbers with the same signs**

       5.   Put +.

       6.   Divide the numbers.

Example 14) Calculate

$(-18) \div (-6) = +3$

Here we put + sign because both 18 and 6 have the same sign which is – then 18 divided by 6 is 3

Example 15) Calculate

$$(+8) \div (+2) = +4$$

Here we put + sign because both 8 and 2 have the same sign which is + then 8 divided by 2 is 4

The following table summarizes the resulting sign if you divide or multiply two numbers:

The sign of the two numbers being multiplied or divided	The resulting sign
Both numbers are negative	Positive
Both numbers are positive	Positive
One number is negative, the other is positive	Negative

Notes about numbers

1) Any number can be divided by 2 without remaining if the last digit of the number is even or/and the last digit of the number is 0, 2,4,6,8.

For example:

0, 12, 134, 7256, and 40118.

Last digit of each of the above numbers is either 0, 2 or 4, or 6, or 8.

2) Any number can be divided by 3 without remaining if the sum of all digits can be divided by 3 without remaining.

For example:

102 can be divided by 3 without remaining because the sum of all digits of 102 is 1 + 0 + 2 = 3 and

521124 can be divided by 3 without remaining because the sum of all digits of 521124 is 5+2+1+1+2+4 =15

3) Any number can be divided by 5 without remaining if the last digit of the number is either 0 or 5

For example:

20, 4785, 35, and 1180.

> **Note:**
>
> Last digit of each of the above numbers is either 0 or 5.

4) Any number can be divided by 4 without remaining if the last two digits can be divided by 4

For example: 24, 116, and 502084.

5) Any number can be divided by10 without remaining if the last digit of the number is 0

For example: 20, 4780, 30, and 1180.

> **Note:**
> Last digit of each of the above numbers is 0

Summary

The four basic mathematical operations are:

> \+ Addition
>
> \- Subtraction
>
> * Multiplication
>
> ÷ Division

- ❖ Adding two (or more) numbers means to find their sum (or total).
- ❖ Subtracting one number from another number is to find the difference between them.
- ❖ Multiplication means times (or repeated addition). A product is the result of the multiplication of two (or more) numbers.
- ❖ Division 'undoes' multiplication.
- ❖ Any number that ends in 0, 2, 4, 6, or 8 can be divided by 2 without remaining.
- ❖ Any number can be divided by 3 without remaining if the sum of all digits can be divided by 3 without remaining.
- ❖ If the last two digits of any number are divisible by 4 without remaining, then the whole number is divisible by 4 without remaining. For example, 1024 is divisible by four because 24 is divisible by 4.
- ❖ Any number that ends in 0 or 5 can be divided by 5 without remaining.
- ❖ Any number whose digits sum up to 9 can be divided by 9 without remaining.
- ❖ Any number that ends in 0 can be divided by 10 without remaining.

Q1) Simplify each of the following:

Hint: The following numbers are having the same sign (negative) so when you add it do the following:

1) Put the same sign (which is negative).

2) Add these numbers.

a) $(-4) + (-5)$

b) $(-9) + (-8)$

c) $(+16) + (+2)$

d) $(+14) + (+12)$

Q2) Simplify each of the following:

Hint: The following numbers are having the different signs so when you add it do the following:

1) Put the sign of the biggest number.

2) Big number – small number.

a) $(+14) + (-15)$

b) $(-19) + (+39)$

c) $(+16) + (-12)$

d) $(-4) + (+12)$

Q3) Simplify each of the following:

Hint: The following numbers are having the same sign (negative) so when you multiply it do the following:

1) Put the positive sign.

2) Multiply these numbers.

a) $(-4) * (-5)$

b) $(-9) * (-8)$

c) $(-6) * (-2)$

d) $(-4) * (-2)$

Section 1.1

Q4) Simplify each of the following:

Hint: The following numbers are having the same sign (positive) so when you multiply it do the following:

1) Put the positive sign.

2) Multiply these numbers.

    a)   (+4) * (+7)

    b)   (+9) * (+8)

    c)   (+6) * (+8)

    d) (+7) * (+8)

Q5) Simplify each of the following:

Hint: The following numbers are having the same sign (negative) so when you multiply it do the following:

1) Put the positive sign.

2) Multiply these numbers.

    a)   (–4) * (–7)

    b)   (–9) * (–8)

    c)   (–6) * (–8)

    d) (–7) * (–8)

Q6) Simplify each of the following:

    a)   (–24) ÷ (–3)

    b)   (+81) * (–9)

    c)   (–46) * (+2)

    d)   (+56) * (+8)

Q7) Subtract 4 from 26.

Q8) Subtract –8 from 25.

Q9) Subtract 5 from –30.

Q10) Subtract –12 from –28.

Section 1.2 Set of Real Numbers

Set: A set is a collection of objects or numbers. Sets are notated by using braces { }.

Elements in a set: the members of a set are called its elements.

For example; if x = {1, 5, 7} we can say x is a set contains 3 elements 1, 5, and 7.

A set that contains no elements is called the empty set (or null set) symbolized by ϕ or { }.

Real numbers consist of all the rational and irrational numbers.

The real number system has many subsets:

1. **Natural Numbers** are the set of counting numbers such as 1, 2, 3, 4, 5, ...
    The three dots after the 5 indicate that the list continues in the same manner without ending.

2. **Whole Numbers** are the set of numbers that include 0 plus the set of natural numbers. {0, 1, 2, 3, 4, ...}.

3. **Integers** are the set of whole numbers and their opposites such as {..., -3, -2, -1, 0, 1, 2,3, ...}.

Rational numbers are any numbers that can be expressed in the form of x/y, where x and y are integers, and $y \neq 0$.

They can always be expressed by using terminating or repeating decimals.

The set of rational numbers is made up of the following:

a. Natural Numbers.

b. Whole Numbers.

c. Integers.

d. Plus, every repeating and terminating decimal.

Terminating decimals are decimals that contain a finite number of digits such as:

2.1, 3.25, 0.125, and 9.25.

> **Note:**
>
> N = Natural numbers = {1,2,3,4,5, ...}
>
> W = Whole numbers = {0, 1,2,3,4,5, ...}
>
> I = Integers = {..., -5, -4, -3, -2, -1, 0, 1,2,3,4,5, ...}

Repeating decimals are decimals that contain an infinite number of digits such as

0.2222... we can write it as $0.\overline{2}$ (you can read it as 0.2 bar). The bar indicates number 2 repeats.

Examples of rational numbers are: $-5, \ 0, \ \dfrac{3}{4}, \ -\dfrac{5}{7}, \ -5.\overline{34}, \ 9.35, \ \text{and} \ -5.2$

Irrational numbers are any numbers that cannot be expressed as x/y (fraction). They are expressed as non-terminating, non-repeating decimals; decimals that go on forever without repeating a pattern such as:

$\pi, 0.12345..., \ \sqrt{2}, -\sqrt{10}, \sqrt{7}, 3.4971...$

Example 1) Classify all the following numbers as natural, whole, integer, rational, or irrational. List all that apply.

  a) 100
  b) 0
  c) -1.41421356237
  d) 1/2
  e) -3

<u>Solution:</u>

  a) 100 is
      1) A natural number.
      2) A whole number.
      3) An integer.
      4) A rational number.

  b) 0 is
      1) A whole number.
      2) An integer.
      3) A rational number.

  c) -1.41421356237 is an irrational number.

  d) ½ is a rational number.

  e) -3 is an integer and a rational number because -3 = -3/1.

> **Note:**
> When taking the square root of any number that is not a perfect square, the resulting decimal will be non-terminating and non-repeating. Therefore, those numbers are always irrational such as $\sqrt{3}$

Example 2) List all numbers from the given set that are: $\{-5, 0, \pi, 3, 4/5, \sqrt{25}\}$

    a. Integers   **b.** Rational numbers   **c.** Irrational numbers   **d.** Real numbers

<u>Solution:</u>

    a. Integers are $\{-5, 0, 3, \sqrt{25}\}$. $\sqrt{25}$ is an integer number because $\sqrt{25} = 5$.

    b. Rational numbers are $\{-5, 0, 3, 4/5, \sqrt{25}\}$.

    c. Irrational number is π

    d. Real numbers are $\{-5, 0, \pi, 3, 4/5, \sqrt{25}\}$.

Example 3) is $\sqrt{49}$ a rational number or irrational number?

<u>Solution:</u>

$\sqrt{49}$ is a rational number because $\sqrt{49} = 7$. You can put $7 = 7/1$ which is a rational number. You can also say that $\sqrt{49}$ is natural, whole, and integer number.

Symbols

$\in$ **used to denote that an element is in a set.**
$\notin$ **is read as it is not an element of a set.**

$$3 \in \{1, 2, 3, 4, 5\}$$

$$p \notin \{a, 5, g, j, q\}$$

p is not an element of $\{a, 5, g, j, q\}$.

What is a set?

A **set** is a group of objects such as numbers, letters, colors, names, etc.

Usually in a math class we will deal with sets of numbers.

Finite sets have an exact number of elements in the set such as {1, 2, 3} or {natural numbers less than 4}.

Infinite Sets have unlimited elements in their set such as {0, 1, 2...} or {whole numbers}.

Subset and Superset

In mathematics, especially in set theory, a set A is a subset of a set B, or equivalently B is a superset of A, if "A" is "contained" inside B, that is all elements of "A" are also elements of B. A and B may coincide. The relationship of one set being a subset of another is called inclusion or sometimes containment.

Example 4) If A and B are sets and every element of A is also an element of B, then:

A is a subset of (or is included in) B, denoted by $A \subseteq B$, we can read it as "A is a subset of B"

or equivalently B is a superset of (or includes) A, denoted by $B \supseteq A$, we can read it as "B is superset of A"

Example 5) The set A = {1, 2} is a subset of B = {1, 2, 3}, thus A ⊆ B

We can say natural numbers are subset of whole numbers.

We can express it as natural numbers ⊆ whole numbers. We can also express

natural numbers ⊆ whole numbers ⊆ integers ⊆ rational numbers.

Venn Diagram

In a Venn diagram, sets are represented by shapes; usually circles, rectangles or ovals. The elements of a set are labelled within the circle. Venn diagrams are especially useful for showing relationships between sets.

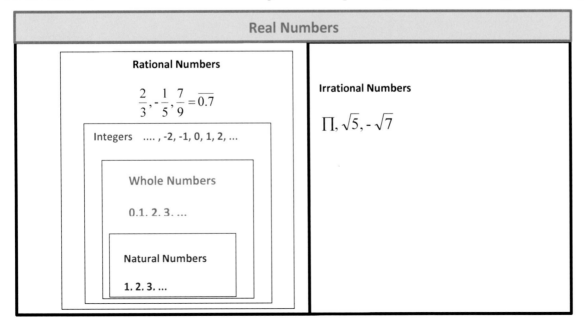

Union and Intersection of Sets

Definition of Union: All the numbers or elements in all the sets.

The union of two sets A, B is denoted by A U B.

A U B = {x| x is in A or x is in B}

Note: The elements of the union are in A or in B or in both. If elements are in both sets, we do not repeat them.

Example 6) If A = {1, 2, 3, 4} and B = {2, 4, 6} what is A U B?

<u>Solution:</u>

A U B = {1, 2, 3, 4, 6}.

Note:

When you find the solution of a union, you can put the elements of the first set (here is A) which is 1, 2, 3, 4 then you will add all elements of B that is not repeatable here in set A. In this case, we only put 6 because 2, and 4 already elements in A.

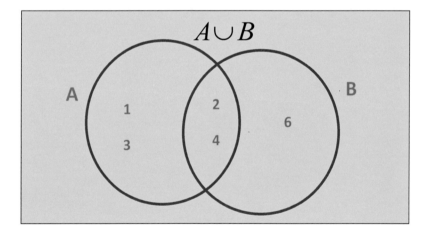

Example 7) If X = {3, 6, 9} and Y = {4, 8, 12} what is A U B?

<u>Solution:</u>

X U Y = {3, 6, 9, 4, 8, 12}

Definition of intersection: The numbers that the sets have in common. In other words, when an element of a set belongs to two or more sets we say the sets will intersect.

The intersection of a set A and a set B is denoted by A ∩ B.

A ∩ B = {x| x is in A and x is in B}.

Example 8) If X = {3, 6, 9} and Y = {4, 8, 12} what is X ∩ Y?

Solution:

X ∩ Y = { } or X ∩ Y = ∅

Notes:

1) Think of this as two events that cannot happen at the same time.

2) {0} is not an empty set because it is not empty! It contains the element zero.

Example 9) If A = {1, 2, 3, 4} and B = {2, 4, 6} what is A ∩ B?

Solution:

A ∩ B = {2, 4} the solution here is 2 and 4 because 2 and 4 exist in both sets (A and B).

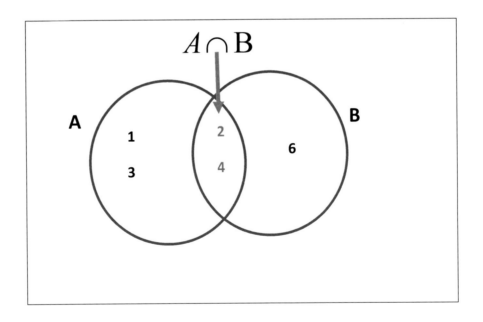

Example 10) If X = {3, 6, 9} and Y = {4, 8, 12} what is X ∩ Y?

Solution:

X ∩ Y = { } or X ∩ Y = ∅

Number Line

A number line is a picture of a straight line on which every point is assumed to correspond to a real number and every real number to a point.

Often the integers are shown as specially-marked points evenly spaced on the line. Although the image below shows only the integers from −9 to 9, the line includes all real numbers, continuing forever in each direction, and numbers not marked that are between the integers.

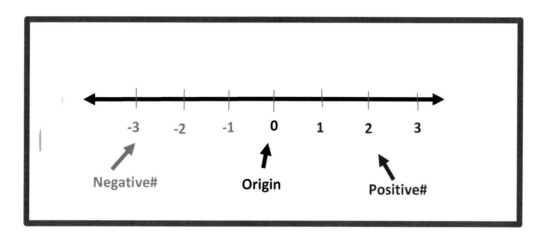

Usually zero in located in the center of the number line we call it the origin. To the right of zero includes all positive real numbers and to the left of zero includes all negative numbers. The number line goes to infinity in both directions.

Example 10) Locate & label the points on the real number line associated with the numbers -2.5, +2.5, 3/4

Solution:

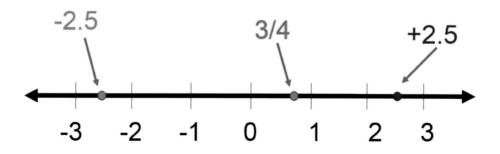

Opposites

Two integers are opposites if they are each in the same distance away from zero, but on opposite sides of the number line. For example, -2.5 is the opposite of +2.5 on the number line (example 10).

Fun Facts about Opposites

✓ Each negative number is the opposite of some positive numbers.

✓ Each positive number is the opposite of some negative numbers for example; - (-a) = a

✓ When you add any two opposites the result is always zero for example; a + (-a) = 0

Example 11) Write a positive or negative integer that describes the following situation.

150 feet below sea level.

Solution:

-150 feet

Absolute Value

The absolute value of a number is its distance from 0 on the number line.

The absolute value of x is written $|x|$.

Note: Absolute value of a number is a distance that cannot be negative. The distance is always positive.

Example 12) Find the value of the followings:

a) $|-7|$

b) $|+8|$

c) $|0|$

Solution

a) $|-7| = 7$ The absolute value of -7 is 7 because -7 is 7 units from 0 on the number line.

b) $|+8| = 8$ The absolute value of +8 is 8 because +8 is 8 units from 0 on the number line.

c) $|0| = 0$ The absolute value of 0 is 0 because 0 is 0 unit from 0 on the number line.

Note: +7 = 7 and +8 = 8.

Example 13) Find the absolute value: $\left| -\sqrt{5} \right|$

Solution:

$\left| -\sqrt{5} \right| = \sqrt{5}.$

Finding the distance between two points on a number line

When we talk about the distance between two points such as x and y on the number line, we are talking about the absolute value of the difference between these two points.
So the distance between x and y on a number line can be found as

$$\left| (x) - (y) \right| \ or \ \left| (y) - (x) \right|$$

Example 14) Tom draws a number line. He plots points at – 4.2 and at 5.7. What is the distance, in units, between Tom's two points?

Solution:

The distance between -4.2 and 5.7 will be

$\left| (-4.2) - (5.7) \right| = \left| -9.9 \right| = 9.9$

Or

$\left| (5.7) - (-4.2) \right| = \left| 5.7 + 4.2 \right| = \left| 9.9 \right| = 9.9$

Example 15) Mary draws a number line. She plots points at -5 and at 10. What is the distance, in units, between Mary's two points?

Solution:

The distance between -5 and 10 will be

$\left| (-5) - (10) \right| = \left| -15 \right| = 15$

The Set of Real Numbers

(Summary)

Sets of Numbers	Definitions
Natural Numbers	All numbers in the set $\{1, 2, 3, 4, \ldots\}$
Whole Numbers	All numbers in the set $\{0,1,2,3,4, \ldots\}$
Integers	All numbers in the set $\{\ldots-3, -2, -1,0,1,2,3, \ldots\}$
Rational Numbers	All numbers x/y such that x and y are integers such as ½, 3/5, -4 $7/9 = 0.\overline{7}$, ...
Irrational Numbers	All numbers whose decimal representation neither terminate nor repeat such as π, $\sqrt{8}$, - $\sqrt{15}$, ...
Real Numbers	All numbers that are rational or irrational

Exercise Set 1.2

Q1) Give an example of a number that is an irrational and real number.

Q2) List all numbers from the given set that are natural numbers, whole numbers, integers, rational numbers, irrational numbers, and real numbers.

    a) $\{9, -45, 0, 3/4, 15, 9.2, 100\}$

    b) $\{8, 0.\overline{7}, 0, - 49, 50\}$

    c) $\{11, -56, 0, 0.65, 50, \pi, 65\}$

    d) $\{-5, \sqrt{7} , 0, 2 ,4\}$

Q3) Give an example of a whole number.

Q4) Give an example of an irrational number.

Q5) Give an example of an integer that is a natural number.

Q6) Give an example of a rational number that is not a natural number.

Q7) Give an example of a number that is an integer, whole number, and natural number.

Q8) Give an example of a number that is a rational number, integer, and real number.

Q9) Give an example of a number that is a whole number, but not natural.

Q10) Are natural numbers subset of whole numbers?

Q11) Lisa draws a number line. She plots points at -15 and at -20. What is the distance, in units, between Mary's two points?

Q12) $\left| -\dfrac{2}{3} \right| = ?$

Q13) If A = {0, 1} and B = {2, 3, 4} what is A U B?

Q14) If A = {0, 5} and B = {5, 13, 14} and C = {5, 10, 15} what is (A U B) U C?

Q15) If X = {5, 6, 7} and Y = {5, 6, 7, 8} what is X ∩ Y?

Section 1.3 Properties of Real Numbers

Properties of real numbers are:

1) Commutative properties
2) Associative properties
3) Distributive properties
4) Identity properties
5) Inverse properties

We are going to define and explain each one of the above properties as in the following:

1) **Commutative Properties**: Changing the order of the numbers in addition or multiplication will not change the result.

   There are two kinds of commutative properties:

   A. Commutative properties of addition: Changing the order of the numbers or elements in addition will not change the result as in the following example:

   a + b = b + a

   When adding two numbers, the order of the numbers does not matter as in the following examples:

   $2 + 3 = 3 + 2$ or $(-5) + 4 = 4 + (-5)$

   B. Commutative properties of multiplication: changing the order of the numbers or elements in multiplication will not change the result as in the following example:

   a * b = b * a

   When multiplying two numbers, the order of the numbers does not matter as in the following example:

   $4 * 5 = 5 * 4$ or $(-3) * 24 = 24 * (-3)$

Example 1) Rewrite the expression 5 *6

using the commutative property.

Solution:

$5 * 6 = 6 * 5$

Example 2) State the property that justifies the following: 4 + x = x + 4.

Commutative property of addition

Example 3) Rewrite the expression using the commutative property.

$$-5 + 7d$$

$$-5 + 7d = 7d - 5$$

2) **Associative properties:** Changing the grouping of the numbers in a sum or multiplication will not change the result.

There are two kinds of associative properties:

 A. Associative properties of addition: Changing the grouping of the numbers in addition or multiplication will not change the result as in the following example:

   $$a + (b + c) = (a + b) + c$$

When adding a group of numbers, the order of the numbers does not matter as in the following example:

   $$3 + (4 + 5) = (3 + 4) + 5 \text{ or } 2 + (-5 + 6) = (2 + -5) + 6$$

 B. Associative properties of multiplication: Changing the grouping of the numbers or elements in multiplication will not change the result as in the following example:

   $$a * (b * c) = (a * b) * c$$

When three numbers are multiplied, it makes no difference which two numbers are multiplied first for examples:

   $$3*(4 * 5) = (3 * 4) * 5 \text{ and } (2 \cdot 3) \cdot 4 = 2 \cdot (3 \cdot 4)$$

Example 4) State the property or properties that justify the following:

$$(8 + 9) + 10 = (8 + 10) + 9$$

Solution:

Commutative and associative

Example 5) Use the associative property to rewrite the following expression then simplify the result.

$5 + (6 + x)$

Solution:

$5 + (6 + x) = (5 + 6) + x$

$\qquad = 11 + x$

Example 6) Rewrite the expression using the associative property 5 * (7d * 8 h).

Solution:

$5 * (7d * 8 h) = (5 * 7d) * 8h$

---

3) **Distributive properties**: Multiplication distributes over addition as in the following example:

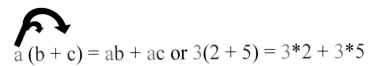

$a (b + c) = ab + ac$ or $3(2 + 5) = 3*2 + 3*5$

Example 7) State the property or properties that justify the following:

$3(x - 10) = 3x - 30$ (Here we multiplied 3 by x and multiplied 3 by -10 which is -30)

Solution:

Distributive property

Example 8) State the property or properties that justify the following:

$-2(-3x - 5) = +6x + 10$ (Here we multiplied -2 by -3x which is +6x and multiplied -2 by -5 which is +10)

Solution:

Distributive property

Example 9) Identify which property justifies the following:

$3(2 + 5) = 3*2 + 3*5$

<u>Solution:</u>

Distributive property

Example 10) Apply the distributive property to the following expression. Simplify when possible

$5(x + 3)$

<u>Solution:</u>

$5 \quad (x + 3) = 5x + 15$

---

4) **Identity properties:** There are two kinds of identity properties:

   A. Additive Identity Property: There exists a unique number 0 such that zero preserves identities under addition as in the following example:

   $a + 0 = a$ and $0 + a = a$

   In other words, adding zero to a number does not change its value.

   B. Multiplicative Identity Property: There exists a unique number 1 such that the number 1 preserves identities under multiplication as in the following example:

   $a * 1 = a$ and $1 * a = a$

Example 11) Identify which property that justifies the followings:

$2 + 0 = 2 \quad$ and $\quad 0 + 2 = 2$

<u>Solution:</u>

Additive Identity Property or we can say Identity Property of Addition.

Example 12) State the property or properties that justify the following:

-25 + 0 = -25

Solution:

Additive Identity Property

Example 13) State the property or properties that justify the following:

$$5\frac{3}{4} + 0 = 5\frac{3}{4}$$

Solution:

Additive Identity Property or we can say Identity Property of Addition.

Example 14) Identify which property that justifies the followings:

2 * 1 = 2   and   1 * 2 = 2

Solution:

Multiplicative Identity Property or we can say Identity Property of Multiplication.

Example 15) State the property or properties that justify the following:

-25 * 1 = -25

Solution:

Multiplicative Identity Property

Example 16) State the property or properties that justify the following:

$$5\frac{2}{7} * 1 = 5\frac{2}{7}$$

Solution:

Multiplicative Identity Property

5) **Inverse properties**: There are two kinds of inverse properties:

   A. Additive Inverse Property: For each real number a there exists a unique real number –a such that their sum is zero (opposites add to zero)
   a + (-a) = 0

   B. Multiplicative Inverse Property: For each real number x, except 0, there exists a unique real number $\frac{1}{x}$ such that their product is one or we can say (the product of a number and its multiplicative inverse (reciprocal) always equals one).

   $$x * \frac{1}{x} = 1$$

Example 17) State the property or properties that justify the following:

   3 + (-3) = 0

Additive Inverse Property or we can write Inverse Property of Addition.

Example 18) State the property or properties that justify the following:
   100(0.01) = 1

Solution:

   Multiplicative inverse property or we can write Inverse Property of Multiplication.

   Note: $0.01 = \frac{1}{100}$ therefore; we can write $100(0.01) = 100 * \frac{1}{100}$ that is equal to 1.

Summary

Properties of Real Numbers

☐ **Commutative**
$$a + b = b + a$$
$$a * b = b * a$$

☐ **Associative**
$$a + (b + c) = (a + b) + c$$
$$a * (b * c) = (a * b) * c$$

☐ **Distributive**
$$a*(b + c) = a*b + a*c$$

☐ **Identity + ×**
$$a + 0 = a \quad \text{and} \quad 0 + a = a$$
$$a * 1 = a \quad \text{and} \quad 1 * a = a$$

☐ **Inverse + ×**
$$a + (-a) = 0$$
$$a * \frac{1}{a} = 1$$

Exercise Set 1.3

Q1) State the property or properties that justify the following:

$$\frac{1}{2}(6x - 14y) = 3x - 7y$$

Q2) Write the letters of the property or properties of the real numbers that justify the expression on the left.

1. xy = yx	**A.** Commutative of addition
2. (7 + a) + 3 = a + (7 + 3)	**B.** Commutative of multiplication
3. 3 + (x + 2) = 3 + (2 + x)	**C.** Associative of addition
4. 5(x + 3) = 5x + 15	**D.** Associative of multiplication
5. 6 + 0 = 6	**E.** Distributive property
6. x*1 = x	**F.** Identity for addition
7. 17 + (-17) = 0	**G.** Identity for multiplication

Q3) Identify which property justifies the following: -5 + 0 = -5?

Q4) Rewrite the expression using the distribution property 8 (4x + 9 y).

Q5) State the property or properties that justify the following: $98 * \dfrac{1}{98} = 1$.

Q6) Rewrite the following expression using the commutative property y * 5x.

Q7) State the property or properties that justify the following: (1 + 2) + 3 = (1 + 3) + 2.

Q8) State the property or properties that justify the following: 1 + [-9 + 3] = [1 + (-9)] + 3.

Q9) Identify which property justifies the following: -3(6) = 6(-3).

Q10) State the property or properties that justify the following: 3*7 − 3*4 = 3(7 − 4).

Section 1.4 Fractions

In mathematics, the set of all numbers (which can be expressed in the form x/y, where x and y are integers and y is not zero). It is called the set of rational numbers (fractions) and is represented by the symbol Q, which stands for quotient. Below is an example of a fraction that contains a numerator in the top of the fraction and denominator at the bottom of the fraction.

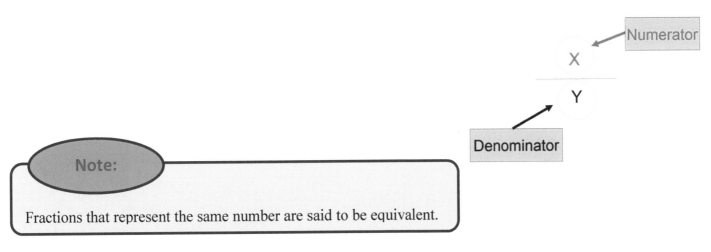

Note:

Fractions that represent the same number are said to be equivalent.

For example: $\dfrac{5}{10} = \dfrac{1}{2}$ and $\dfrac{2}{8} = \dfrac{1}{4}$

Example 1) Write 3/5 as an equivalent fraction with denominator 10.

Solution:

$$\frac{3}{5} = \frac{3}{5} * \frac{2}{2} = \frac{6}{10}$$

Reciprocals

Two numbers whose product is 1 are reciprocals of each other.

Examples of Reciprocals:

$\dfrac{3}{4}$ and $\dfrac{4}{3}$

$\dfrac{1}{5}$ and 5

$-\dfrac{1}{7}$ and -7

Least Common Denominator

The least common denominator (LCD) is the smallest number divisible by all the denominators.

It is also called as Least Common Multiple (**LCM**).

Where:

L for the smallest number

C for all terms in common

M from their list of multiples

There are two methods for finding the Common Multiples of two or more numbers…

Method# 1) Use Multiple Lists Method, for example, the first five multiples of 6 are 6, 12, 18, 24, 30.

Method# 2) Use Prime Factorization Method, for example, the prime factorization of 12 is $12 = 2*2*3$ where 2, 2, and 3 are the prime factors of 12.

Note: A multiple of a number is the product of the number and any nonzero whole number.

Example 2) Find the LCM of 4 and 9 by using the multiple lists method (method #1).

Solution:

Step 1: Create a list of multiples for each number:

4: 4, 8, 12, 16, 20, 24, 28, 32, 36, 40, 44, 48…

9: 9, 18, 27, 36, 45, 54, 63…

Step 2: Circle the first multiple the numbers have in common.

4: 4, 8, 12, 16, 20, 24, 28, 32, ⃝36 40, 44, 48…

9: 9, 18, 27, ⃝36 45, 54, 63…

The LCM of 4 and 9 is 36.

Here 36 is the smallest number in the above lists that can be divided by 4 and 9 without remaining.

Example 3) Find the LCM of **4** and **9** by using the prime factorization (method #2).

Solution:

Step 1: Find the prime factorization of each number.

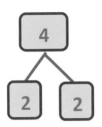   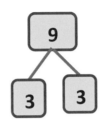

$4 = 2 * 2 = 2^2$ (Here the prime factors for 4 are 2 &2) and $9 = 3*3 = 3^2$ (here the prime factors for 9 are 3 & 3).

Step 2: To find the LCM, multiply each prime factor once with the greatest exponent used in any of the prime factorizations.

$LCM = 2^2 * 3^2 = 4 * 9 = 36$

Example 4) Find the LCD (LCM) of each of the following fraction's denominator $\frac{2}{3}$ and $\frac{5}{4}$.

Solution:

LCD = 12 because 12 is the smallest number into which 3 and 4 will both divide without remaining.

Example 5) Find the LCD of each of the following fraction's denominator $\frac{2}{5}$ and $\frac{7}{10}$.

Solution:

LCD = 10 because 10 is the smallest number into which 5 and 10 will both divide without remaining.

Note:

- The LCD between two numbers (big and small) is equal to the big number if the big number can be divided by the small number without remaining for example; the LCD between 9 & 18 is 18 because 18 can be divided by 9 without remaining.
- If the big number cannot be divided by the small number without remaining, use either method 1 or 2 explained (Page# 30).

Example 6) Find the least common multiple (LCM) of 4, 6, and 15.

<u>Solution:</u>

    1) Write the prime factorization of each number in exponential form.

$4 = 2 * 2 = 2^2$

$6 = 2 * 3$

$15 = 3 * 5$

    2) To find the LCM, multiply each prime factor once with the greatest exponent used in any of the prime factorizations.

    3) LCM $= 2^2 * 3 * 5$ (we did choose 2^2 not 2 because 2^2 has higher exponent).

Comparing Two Fractions

Students need to be able to determine if two fractions are equal or if one fraction is greater or less than the other.

Students will learn to correctly place >, <, or = between two fractions.

    > is the greater than symbol

    < is the less than symbol

Butter Fly Method

Comparing fractions and finding which unlike fraction (different denominators) is larger or smaller. For example:

1. Write the two fractions such as $\frac{2}{3}$ and $\frac{3}{5}$.

2. Multiply the first numerator (2) with the second denominator (5).

3. Write the product above the first fraction. It will be $2*5 = 10$.

4. Multiply the second numerator (3) with the first denominator (3).

5. Write the product above the second fraction. It will be $3*3 = 9$.

6. Decide which product is greater and put the greater than or less than symbol between the two fractions.

So, to compare two fractions such as 2/3 and 3/5 you can use the previous 6 steps which is the "Butterfly" method to know which one of these fraction is bigger.

10   **>**   9

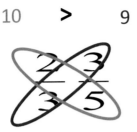

The greater product is written above; the greater fraction is.

Thus, $\dfrac{2}{3} > \dfrac{3}{5}$ or we can write $\dfrac{3}{5} < \dfrac{2}{3}$

Mixed Numbers: A mixed number has a part that is a whole number such as 1 and a part that is a fraction such as ¾.

 = $1\dfrac{3}{4}$

Another example of a mixed number is

$2 + \dfrac{3}{4} = 2\dfrac{3}{4}$ In this example, the whole number is 2 and the fraction is ¾.

Improper fraction – a fraction with a value greater than or equal to 1 for example; 5/4, 6/3, 9/2, etc.

Converting a Mixed Number to an Improper Fraction

To convert a mixed number such as $2\dfrac{3}{4}$ to an improper fraction, do the following:

1) Multiply the denominator (which is 4 in this example) of the fraction by the whole number (which is 2) and add the numerator (3) to this product.
2) Place the result from step 1 over the denominator in the original mixed number.

Thus,

$$2\dfrac{3}{4} = \dfrac{4 * 2 + 3}{4} = \dfrac{8 + 3}{4} = \dfrac{11}{4}\ .$$

Converting an Improper Fraction to a Mixed Number

To convert an improper fraction such as 15/2, to a mixed number, do the following:

1) Divide the denominator into the numerator. Record the quotient and the remainder.
2) Put the remainder over the original denominator.
3) Write the mixed number using the form:

$$quotient \ \frac{remainder}{original \ denominator}$$

Example 7) Convert 15/2 to a mixed number.

Solution:

$\dfrac{15}{2}$ 15 divided by 2 yields 7 with a remainder of 1.

$$\frac{15}{2} = 7\frac{1}{2}.$$

Note:

➢ If there is no remainder, then your answer is a whole number!

For example:

$\dfrac{10}{2}$ 10 divided by 2 yields 5 with no remainder.

$$\frac{10}{2} = 5.$$

Example 8) Express the improper fraction $\dfrac{19}{3}$ as a mixed number.

Solution:

$$\frac{19}{3} = 6\frac{1}{3}.$$

Adding and Subtracting Fractions with the same Denominator

When you add, and subtract fractions, you COPY the denominator, then you work on the top. Remember, you copy the denominator and then you work on the numerators.

$$\frac{2}{5} + \frac{1}{5} = \frac{\square}{\square}$$

Look at the above example. What is the denominator?

The denominator is 5.

What do you do with the denominator?

You copy it in the answer so the denominator is 5.

$$\frac{2}{5} + \frac{1}{5} = \frac{\square}{5}$$

Now we can add the numerators (numbers on the top). What do we get when we add 2 + 1?

2 + 1 = 3.

We put the 3 on top in the answer so,

$$\frac{2}{5} + \frac{1}{5} = \frac{3}{5}$$

Example 9) Simplify the following:

$$\frac{7}{9} - \frac{2}{9}$$

Here it is a different example. First look at the sign. We are subtracting in this example.

Next look at the denominators. What do we do with the denominator?

The denominator is 9.

On the top the sign tells us to subtract 7 – 2, what's the answer?

The answer is 7 – 2 = 5. We put the 5 on top. Therefore,

$$\frac{7}{9} - \frac{2}{9} = \frac{5}{9}$$

Note: To add or subtract fractions with like denominators

- ➢ If the denominators are already the same, then you're ready to add or subtract.
- ➢ Add or subtract the numerators. Keep the denominators the same.
- ➢ Simplify (reduce) the fraction, if possible.
- ➢ If you end up with an improper fraction, change it to a mixed number.

Example 10) Simplify the following:

$$\frac{32}{45} - \frac{29}{45}$$

<u>Solution</u>

$$\frac{32}{45} - \frac{29}{45} = \frac{3}{45} = \frac{1}{15}$$

Here after we divide both top and bottom of the fraction by 3, we get 1/15

Adding and Subtracting Unlike Fractions

To add (or subtract) unlike two fractions such as 1/5 and 2/3, rewrite them as equivalent fractions with a common denominator by doing the followings:

1. List the multiples of both denominators.

   Multiples of 1/5: 5, 10, (15), 20, ...

   Multiples of 2/3: 3, 6, 9, 12, (15), 18, ...

2. Find the least common multiple (LCM). Here LCM = 15.

3. Write new fractions with the LCM as the new denominator.

4. Find the factor you multiply by to get from your original denominator to your new denominator.

5. Use that same factor, and multiply it by your original numerator to get a new numerator.

$$\frac{1}{5} = \frac{1}{5} * \frac{3}{3} = \frac{3}{15}.$$

$$\frac{2}{3} = \frac{2}{3} * \frac{5}{5} = \frac{10}{15}.$$

Here we multiplied 1/5 by 3/3 so that we change the denominator to 15 and we multiplied 2/3 by 5/5 so that we change the denominator to 15 (we rewrite both fractions as equivalent fractions with a common denominator 15). Thus,

$$\frac{3}{15} + \frac{10}{15} = \frac{13}{15}.$$

Example 11) Subtract

$$\frac{3}{8} - \frac{1}{5}$$

The LCM between 8 & 5 = 40

Solution:

The LCM between 8 & 5 is 40.

$$\frac{3}{8} - \frac{1}{5} = \frac{3}{8} * \frac{5}{5} - \frac{1}{5} * \frac{8}{8} =$$

$$\frac{15}{40} - \frac{8}{40} = \frac{15 - 8}{40} = \frac{7}{40}.$$

Example 12) Tony made a pizza with chicken covering 1/2 of the pizza and mushrooms covering 1/3.

What fraction of the pizza is covered by chicken and mushrooms?

Solution:

You can add 1/2 and 1/3, which are unlike fractions. The LCM between the dominator 2 and 3 is 6.

$$\frac{1}{2} + \frac{1}{3} = \frac{1}{2} * \frac{3}{3} + \frac{1}{3} * \frac{2}{2} =$$

$$\frac{3}{6} + \frac{2}{6} = \frac{5}{6}.$$

Multiplying Fractions

To multiply fractions, simply multiply the two numerators and the two denominators.

Example 13) Multiply

$$\frac{3}{4} * \frac{2}{5} = \frac{3*2}{4*5} = \frac{3*2}{4*5}$$

Note: When multiplying fractions, we can simplify the fractions and simplify diagonally.

Example 14) Multiply and simplify the following:

$$\frac{3}{8} * \frac{7}{9}$$

Solution:

$$\frac{3}{8} * \frac{7}{9} = \frac{\overset{7}{\cancel{21}}}{\underset{24}{\cancel{72}}} = \frac{7}{24}$$

Hint: We divide both 21 and 72 by 3.

Note: Before you multiply, you can simplify the numerator of any denominator if possible as in the following example:

Example 15) Multiply and simplify the following:

$$\frac{2}{9} * \frac{(-7)}{(4)}$$

<u>Solution:</u>

$$\frac{\overset{1}{\cancel{2}}}{9} * \frac{-7}{\underset{2}{\cancel{4}}} = \frac{1}{9} * \frac{-7}{2} = \frac{-7}{18}$$

Dividing Fractions

To divide two fractions, use the **KCF** Method where;

K for keep (keep the first fraction).

C for change (change the operation to multiply).

F for flip (flip the 2nd fraction).

Example 16) Divide and simplify the following:

$$\frac{2}{9} \div \frac{5}{7}$$

<u>Solution:</u>

Change operation

$$\frac{2}{9} \div \frac{5}{7} = \frac{2}{9} * \frac{7}{5} = \frac{14}{45}$$

Flip the 2nd fraction

Example 17) Simplify the following:

$$\frac{2}{9} \div \frac{8}{3}$$

Solution:

$$\frac{2}{9} \div \frac{8}{3} = \frac{\overset{1}{\cancel{2}}}{\underset{3}{\cancel{9}}} * \frac{\overset{1}{\cancel{3}}}{\underset{4}{\cancel{8}}} = \frac{1*1}{3*4} = \frac{1}{12}$$

Exercise Set 1.4

Q1) Write 7/25 as an equivalent fraction with denominator 100.

Q2) Find the LCM of 8 and 9 by using the multiple lists method (method #1).

Q3) Find the LCM of 10 and 12.

Q4) Find the LCM of 16 and 12.

Q5) Jim made a pizza with chicken covering 1/3 of the pizza and vegetables covering 1/4.

What fraction of the pizza is covered by chicken and vegetables?

Q6) The Pacific Ocean covers 1/3 of Earth's surface. The Atlantic Ocean covers 1/5 of Earth's surface. Find the fraction of Earth's surface that is covered by both oceans.

Q7) Simplify the following: $\dfrac{4}{7} \div \dfrac{8}{14}$.

Q8) Multiply $\dfrac{-16}{5} * \dfrac{-8}{34}$.

Q9) Subtract $\dfrac{43}{8} - \dfrac{22}{8}$.

Q10) Add $\dfrac{3}{8} + \dfrac{2}{5}$.

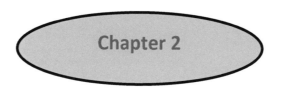

Chapter 2

Variables, Linear Equations, Word Problems, and Inequalities

Section 2.1 Variables & Translating English Phrases and Algebraic Expressions

Vocabulary Definitions

A **variable** is a letter used to represent an unknown value.

For example: a, b, c, …x, y, z

A **constant** is a value that does not change for example; numbers such as: 2, 8, 29, -6, …etc.

> **Note:** Constants are fixed values such as numbers, letters, and strings.

A **numerical expression** includes numbers and operations.

For example: 5+3, 8, …

An **algebraic expression** contains variables, numbers and operations.

> **Note**: No equal sign in any algebraic expression.

For example: x + 3 - 2, a + 2.5b, 5x + 1000

An **Equation** is a mathematical statement that two expresses are equal.

For example: x + 1 = 4, 2 = 5x, 3n + 5 = 11

A **term** is a constant, variable, the product or quotient of constants and/or variables for example: 5, m, 5m, 6/m.

Example 1) Evaluate the algebraic expression x + 6 when x = 5.

<u>Solution:</u>

x + 6 =
↓
5 + 6 (here we substitute 5 for x).

= 11 (here we add 5 to 6 which is 11).

Example 2) Evaluate the algebraic expression $4(x) + 5$ when $x = 2$.

Solution:

$4(x) + 5 =$

$4(2) + 5$ (Here we substitute 2 for x).

$= 8 + 5$ (Here we multiply 4 by 2 which is 8).

$= 13$ (Here we add 8 to 5 which is 13).

Example 3) Identify the following for the equation $5x - 9 = 1$:

    1) Variable **2)** Constants **3)** Expressions **4)** Equation **5)** Terms.

Solution:

1.) Variable	X
2.) Constants	5, -9, 1
3.) Expressions	5x -9, 1
4.) Equation	5x -9= 1
5.) Terms	5x, -9, 1

Example 4) Identify the following for the equation $6a - 9b = 7$:

1) Variable   **2) Constants**   **3) Expressions**   **4) Equation**   **5) Terms**

Solution:

1.) Variables	a,b
2.) Constants	6, -9, 7
3.) Expressions	6a -9b, 7
4.) Equation	6a -9b= 7
5.) Terms	6a,-9b, 7

47

Translating English Phrases and Algebraic Expressions

To translate an English phrase (verbal statement) into a mathematical (variable) expression or equation, it requires recognizing the verbal phrases that translate into mathematical operations. The following charts contains some key words or phrases that may help students to translate these English words and phrases to algebraic expressions:

Chart# 1

Addition +	Subtraction −	Multiplication *	Division ÷
add	subtract	multiply	divide
sum	difference	product	quotient
plus	minus	times	ratio
more than or greater than	less than	twice or double	split equally
increased by	decrease by	of or each	to
total	less	by	per

Chart# 2

English Phrases or words	Mathematical Expression
Different of two numbers	X − Y
Sum of two numbers	X + Y
is	equals =
A number (or what)	X
Twice a number	2X
Four more than twice x	2x + 4
Of or times	Multiply
Product	Multiply
To	Division (/)
Three times a number	3x
Six less than twice x	2x -6
4%	4/100=0.04
4.40%	4.4/100 = 0.044
At most	≤
At least	≥

Chart# 3

English Phrases	Mathematical Expression
4 more than x	x + 4
Seven subtracted from twice x	2x – 7
4 less than x gives 1	x – 4 = 1
The difference between 15 and a number is 10	15 – x = 10
The sum of X and 7	X + 7

Below are more examples of English phrases and their mathematical expression:

Chart# 4

English Phrases	Mathematical Expression
The sum of 4 and 5 is 9	4 + 5 = 9
The difference of 9 and 2 is less than 7	9 – 2 < 7
The product of 5 and x is greater than 10	5*x > 10
The quotient of y and 2 is equal to the sum of Y and 5	Y/2 = Y + 5
Three to the second power (Base 3, exponent 2)	3 ^ 2 = 3 * 3 = 9
Two to the fifth power (Base 2, exponent 5)	2 ^ 5 = 2.2.2.2.2 = 32

Example 5) Rewrite the following verbal expression as algebraic expression:

The sum of S and 12

<u>Solution:</u>

S +12

Example 6) Rewrite the following verbal expression as algebraic expression:

The product of 15 and a number.

<u>Solution:</u>

15 * x = 15x

Example 7) Rewrite the following verbal expression as algebraic expression:

n to the seventh power.

Solution:

$$n^7$$

Example 8) Rewrite the following verbal expression as algebraic expression:

56 increased by twice a number.

Solution:

2x + 56

Example 9) Rewrite the following verbal expression as algebraic expression:

Five less than a number

Solution:

x – 5

Example 10) Rewrite the following verbal expression as algebraic expression:

Forty-three less than n is equal to 35.

Solution:

n – 43 = 35

Example 11) Rewrite the following verbal expression as algebraic expression:

Seven more than y is 13

<u>Solution:</u>

$y + 7 = 13$

Example 12) Find the product of 5 and y is greater than or equal to the difference of y and 17

<u>Solution:</u>

$5y \geq y - 17$

Note: Multiplication and division of variables can be written in several ways, as shown in the table below:

Multiplication	Division
ab	a/b
a(b)	a ÷ b
a . b	$\dfrac{a}{b}$
a * b	
(a)(b)	
(a)b	
a x b	

Q1) Evaluate the algebraic expression 3(x) - 9 when x = 5.

Q2) Identify the following for the equation -4x – 9 = -29

1) Variable, 2) Constants, 3) Expressions, 4) Equation, 5) Terms.

Q3) Rewrite the following verbal expression as algebraic expression:

The difference of x and 4 is 8.

Q4) If 36 is subtracted from the square of a certain number x the result is 14. Which of the following equations determines the correct value of b?

$a)\ x^2 - 36 = 14$ $b)\ x^2 = 36$ $c)\ x - 36 = 196$ $d)\ 36 - x^2 = 14$

Q5) Evaluate the following algebraic expression for the given value of the variables.

12(x) + y for x = 7 and y = 4.

Q6) Jeff types 45 words per minute. Write an expression for the number of words he types in "m" minutes.

Q7) Kevin earns $5 for each car he washes. Write an expression for the number of cars Kevin must wash to earn "D" dollars.

Q8) Choose the correct answer for the following algebraic expression:

Two times the quantity of a number plus three.

    a) 3(n + 2)   b) 2(n +3)   c) 3n – 2   d) 2(n –3)   e) 2n – 3

Q9) Write an Algebraic Expression for the following Situation:

Tom's brother is 5 years younger than Tom.

Q10) Evaluate $\dfrac{18}{a}$ + 4b, for a = 3 and b = 2.

Section 2.2 Like Terms and Unlike Terms

Like Terms: Two or more terms with the same variable parts are called similar (or like) terms. For examples; all the following terms are like terms: 3x, -5x, $\frac{4}{7}x$.

Definition:

Term is a number (coefficient) multiplied by a letter (**variable**) raised to an **exponent** (power) for example, $3x^5$ is a term where,

3 is the coefficient, **x is a variable**, and **5 is the exponent.**

Like terms have the same variable part raised to the same power and same exponents as in the following:

1) 5x and 8x are like terms. It has the same variable part which is x raised to the same power which is 1.

2) $\frac{2}{3}x^5$ and $4x^5$ are like terms. It has the same variable part which is x and raised to the same power which is 5.

3) $8xyz^2$ and $-5xyz^2$ are like terms because they have the same variables and powers.

Unlike terms: Two or more terms that are not like terms, i.e. they do not have the same variables or powers as in the following examples:

1) 3abc and 3xyz are unlike terms because they have different variables.

2) 5x and 5y are unlike terms because they have different variables.

3) $\frac{2}{3}x^2$ and $4x^7$ are unlike terms because they have different powers.

Combining Like Terms

Below are some notes about combining like terms:

1. We can only add or subtract like terms. On a table, we have 4 apples and 2 bananas. We cannot add the 4 apples to the 2 bananas - they are not the same kind of objects.

2. To combine 2x and 3x, think about adding 2 apples and 3 apples which is 5 apples. Thus, to add 2x to 3x, the result is 5x.

3. You can also combine numbers that have no variables because they are constants.

4. When an operation separates an algebraic expression into parts, each part is called a term. Example: $2x + 5$ (2x and 5 are both terms).

5. When adding, or subtracting like terms, keep the same variables.

Example 1) Simplify the following:

$3x + 5 + 2x - 3$.

Solution:

$3x + 5 + 2x - 3 = 5x + 2$ (Here we add $3x$ to $2x$ which is $5x$, and add 5 to -3 which is +2).

Example 2) Simplify the following expression:

$5(2x - 8) - 3$.

Solution:

$5(2x - 8) - 3 = 10x - 40 - 3$ (Here we use the distributive property). $5(2x - 8) = 10x - 40$.

$= 10x - 43$ (Here we add -40 to -3 which is -43).

Example 3) Determine the sum $(-50X) + (-5X)$.

Solution:

$(-50X) + (-5X) = -55x$.

Example 4) Simplify the following expression:

$2x - 3y + 3x - 4y$.

Solution:

$2x - 3y + 3x - 4y = 5x - 7y$.

Example 5) Simplify the following expression:

$5(x - 2) - (3x + 4)$

Solution:

$5(x - 2) - (3x + 4)$

$= 5x - 10 - 3x - 4$

$= 2x - 14.$

Exercise Set 2.2

Q1) Simplify the following expression:

$5x - 4x.$

Q2) Simplify the following expression using distributive property then combine the like terms:

$5 - 2(a + 1).$

Q3) Simplify the following expression using distributive property then combine the like terms:

$-9(2x + 1) - (x + 5).$

Q4) Combine the like terms then Find the value of the following expression when x is -5:

$7x - 4 - x - 3.$

Q5) Simplify the following:

$2y^2 + 6y^2 - 10y^2 + 3y.$

Definition:

Any equation that can be put in the standard form $ax + by = c$.

Where a, b, and c are real numbers and **a** and **b** are both not 0, is called a linear equation in two variables.

Properties of Equality

- ➢ Addition property of equality: You can add the same term to both sides of an equation.
- ➢ Subtraction property of equality: You can subtract the same term from both sides of an equation.
- ➢ Multiplication property of equality: You can multiply both sides of an equation by the same term.
- ➢ Division property of equality: You can divide both sides of an equation by the same term.

Notes:

1. Whatever you do to one side of an equation, you MUST do to the other side!
2. When you move one term from one side to the other side, you must change the sign to its opposite sign for example:

$$x - 2 = 8$$
$$x = 8 + 2$$

Strategy for Solving Linear Equations in One Variable

Step 1) Use the distributive property to separate terms, if necessary for example:

$$3(x + 2) = 15$$

$$3x + 6 = 15$$

Step 2) If fractions are present, consider multiplying by the LCD to clear fractions for example:

$$\frac{1}{3}x + \frac{1}{2}x = 5$$

$6(\frac{1}{3}x + \frac{1}{2}x) = 6*(5)$. Here we multiplied both sides by 6 because 6 is

the LCD of 2 & 3 to clear the fractions.

$$2x + 3x = 30$$

Step 3) Combine similar terms on each side of the equation.

$$2x + 3x = 30$$

$$5x = 30$$

Step 4) Use the addition property of equality to get all variable terms on one side of the equation and all constant terms on the other side for example:

5a + 7 = -2a + 42 (to move any term from one side to another, change it to its opposite sign).

5a + 2a = 42 – 7   (Here we move -2a from right to the left of the equation, we change to +2a and we move

+7 from left to right, we change to -7).

7a = 35   (Here we combine the like terms 5a+2a is 7 and 42-7=35).

Step 5) Solve for that variable (find the solution):

7a = 35   (We usually divide both sides to whatever number next to the variable (coefficient number).

$\dfrac{7a}{7} = \dfrac{35}{7}$ Here we divide both side by 7 because 7 is the coefficient number.

a = 5

Step 6) Check your solution in the original equation.

The solution was **a = 5**. We are going to plug **5** in the original equation which is:

5(**a**) + 7 = -2(**a**) + 42

5(**5**) + 7 = -2(**5**) + 42

25+7 = -10 +42

32 = 32. (This is a true statement, therefore; our solution is correct).

Note
Anytime you see the word **solve**, it means find the solution to the variable in the equation.

Example 1) Solve the following equation:

$$5(x - 3) + 2 = 5(2x - 8) - 3$$

Solution:

$5(x - 3) + 2 = 5(2x - 8) - 3$

$5x - 15 + 2 = 10x - 40 - 3$

$-15 + 2 + 40 + 3 = 10x - 5x$

$+30 = 5x$

$$\frac{30}{5} = \frac{5x}{5}$$

$6 = x$

Example 2) Simplify the following expression:

$x - 1.2 = -8.9$

Solution:

$x - 1.2 = -8.9$

$x = -8.9 + 1.2$

$x = -7.7$

Example 3) Solve the following equation:

$3x - 2 = 7$

Solution:

$3x - 2 = 7$

$3x = 7 + 2$

$3x = 9$

$$\frac{3x}{3} = \frac{9}{3}$$

$x = 3$

The Cartesian Coordinate System

The Cartesian coordinate system was named after René Descartes. It consists of two real number lines, the horizontal axis (x-axis) and the vertical axis (y-axis) which meet in the right angle at a point called the origin. The two number lines divide the plane into four areas called quadrants.

The quadrants are numbered using Roman numerals as shown below. Each point in the plane corresponds to one and only one ordered pair of numbers (x, y).

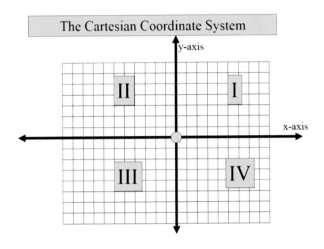

Notes:

1. A solution of an equation in two variables is an ordered pair of real numbers that satisfy the equation.
   For example, (4, 3) is a solution of 3x - 2y = 6.

2. The solution set of an equation in two variables is the set of all solutions of the equation.

3. The graph of any linear equation is a straight line.

Ordered Pairs

Ordered Pair is a pair of numbers that can be used to locate a point on a coordinate plane. For example: (3, 5) is an order pair where 3 is the x-coordinate, 5 is the y-coordinate.

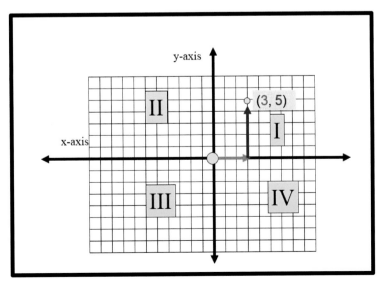

The following is the graph of the order pairs **(2, 2), (- 4, 3), (-3,-3), and (4,-2)**:

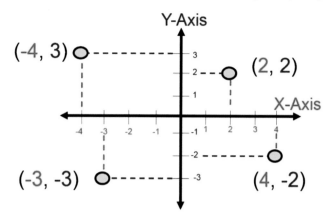

Steps for Graphing Linear Equation

Step1: Find any two ordered pairs. This can be done by using a convenient number for one variable and solving for the other variable.

Step 2: Graph the two ordered pairs found in step 1.

Step 3: Draw a straight line through the two points graphed in step 2.

Example 4) Graph y = -2x + 1.

<u>Solution:</u>

Let us choose two convenient numbers for x such as 0 and 1 (we can choose any other two numbers).

To find the first order pair, plug **x = 0** in the equation **y = -2x +1**.

$$y = -2(0) +1= 0+1= 1 \text{ thus, the first order pair is } (0, 1).$$

To find the second order pair, plug **x=1** in the equation **y = -2x +1**.

$$y = -2(1) +1= -2+1= -1 \text{ thus, the second order pair is } (1, -1).$$

Now all we need, draw both pairs (0, 1) and (1, -1) then connect them with the straight line as shown below:

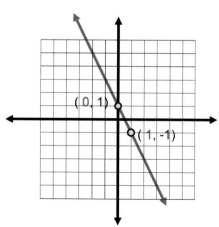

X-Intercept and *Y*-Intercept

The x-intercept of a straight line is the x-coordinate of the point where the graph crosses the x-axis and the y-intercept of a straight line is the y-coordinate of the point where the graph crosses the y-axis as shown below:

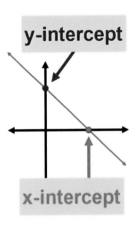

1. To find the x-intercept, substitute it with 0 for y and solve for x.
2. To find the y-intercept, substitute it with 0 for x and solve for y.
3. If you have x-intercept, x has a value and y is zero.
4. If you have y-intercept, y has a value and x is zero.

Example 5) Find the x-intercept and the y-intercept for $x + y = 1$.

<u>Solution:</u>

To find the x-intercept, substitute it with 0 for y and solve for x.

$x + y = 1$

$x + 0 = 1$

$x = 1$

Thus, x-intercept is $(1, 0)$

To find the y-intercept, substitute in 0 for x and solve for y.

$x + y = 1$

$0 + y = 1$

$y = 1$

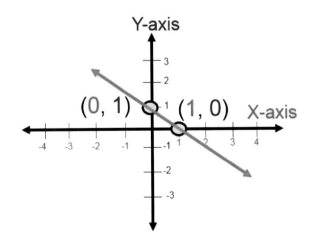

Thus, y-intercept is $(0, 1)$.

Example 6) Find the x-and y-intercepts for $2x + y = 2$.

Solution:

To find x-intercept, let $y = 0$.

$2x + y = 2$

$2x + 0 = 2$

$2x = 2$

$x = 1$

Thus, x-intercept is $(1, 0)$.

To find y-intercept, let $x = 0$.

$2x + y = 2$

$2(0) + y = 2$

$y = 2$

Thus, y-intercept is $(0, 2)$.

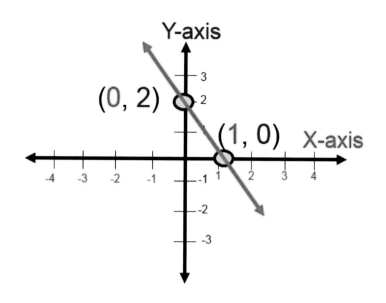

Slope of a Line

Slope is often used to describe the measurement of the steepness, incline, gradient, or grade of a straight line as shown below:

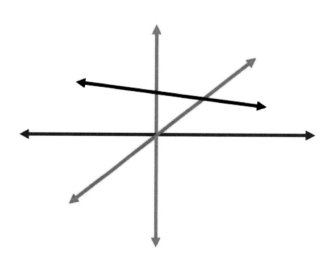

62

Notes:

1. Slope describes the direction of a line.
2. A higher slope value indicates a steeper incline.

3. If $x_1 \neq x2$, the slope of the line through the distinct points (x_1, y_1) and (x_2, y_2) is

$$m = \frac{(y_2) - (y_1)}{(x_2) - (x_1)}.$$

Example 7) Find the slope between (-3, 6) and (5, 2).

<u>Solution:</u>

$$m = \frac{(y2) - (y1)}{(x2) - (x1)} = \frac{(2) - (6)}{(5) - (-3)} = \frac{-4}{8} = \frac{-1}{2}$$

The Slope-Intercept Form:

The equation of the Slope-Intercept Form is **y = mx + b** where,

m represents the **slope**.

b represents the **y-intercept**.

Important Notes

❖ The equation y = mx +b is called the slope intercept form for a line.

❖ The graph of this equation (linear equation) is a straight line.

❖ m is the slope.

❖ The line crosses the y-axis at b.

❖ The point where the line crosses the y-axis is called the y-intercept.

❖ The (x, y) coordinates for the y-intercept are (0, b). This means y-intercept = (0, b).

❖ b is the y-intercept (this means y-intercept = b).

❖ To write an equation, you need to know:

   1) Slope (m).

   2) y-intercept (b).

63

Example 8) Write the equation of a line that has a y-intercept of -3 and a slope of -4.

Solution:

$m = -4$

$b = -3$

$y = mx + b$

$y = -4x + (-3)$

$y = -4x - 3$

Example 9) Find the slope and y-intercept of the line $y = 2x + 5$.

Solution:

By comparing $y = 2x + 5$ to

$$y = mx + b$$

We can figure it out that:

$m = 2$

Y-intercept $= b = 5$.

Example 10) Write an equation in slope-intercept form for the line that satisfies the following condition. The slope is 5 and passes through (2, 12).

Solution:

$y = mx + b$

$m = 5, \quad x = 2, \quad y = 12$.

To find b, substitute the numbers for m, x, and y.

$12 = (5)(2) + b$

$12 = 10 + b$

$12 - 10 = b$

$2 = b$

Slope-Intercept Form is $y = mx + b$.

$$y = 5x + 2.$$

> **Hint**
>
> To write an equation, you need two things:
>
> 1) slope (m)
> 2) y – intercept (b)

Example 11) What is the equation of the line that contains the points (-1, 1) and (2, 7)?

To find the equation of the line, we need to find m & b then plug both m & b in the equation y = mx + b.

1) Calculate the slope m.

$$m = \frac{y_2 - y_1}{x_2 - x_1} = \frac{7-1}{2-(-1)} = \frac{6}{3} = 2.$$

2) Find b by plugging:

    1) y = 7

    2) m = 2

    3) x = 2.

In the slope intercept form:

y = mx + b
7 = 2*2 + b
7 = 4 + b
3 = b

3) Plug the values of m=2 and b= 3 (found in step 1 and 2) in the slope intercept form y = mx + b.

$$y = mx + b$$
$$y = 2x + 3$$

Example 12) Write equation in slope-intercept form then find the slope and the y-intercept for the following:

$$y - 3x + 4 = 0$$

y − 3x + 4 = 0
y = +3x − 4
y = mx + b
m = 3

Y-Intercept = b= -4.

Q1) Solve the following equation: $2x - 1 = 9$.

Q2) Find the X and Y intercepts for the following linear equation then graph it: $2x + 3y = 6$.

Q3) Find the slope and y-intercept of the line $y = 2x - 7$.

Q4) Find the slope of the lines that contain the following two points: (3, 4) and (-6, -2).

Q5) Find the slope of the following equation: $y = -5x + 6$.

Q6) Find the slope and y-intercept of the following equation: $8x + 11y = 7$.

Q7) Find an equation in slope-intercept form for the line passing through the point (0,2) with m=1/3.

Q8) Write the equation of a line that has a y-intercept of 15 and a slope of 14.

Q9) Solve the following equation for x:

$$3ax = \frac{1}{5}.$$

Q10) Graph the following order pairs (2, 1), (- 4, 0), (0, 3), and (-5, -1).

Q11) Which equation represents the line that passes through (1, - 3), (- 2, 3)?

Q12) Find the slope and y-intercept for the line $2y + 2 = 4x$.

Q13) Find an equation in slop-intercept form for the line passing through the point (0, 2) with $m = 3$.

Q14) Which equation best describes the data in this table?

x	-3	0	3
y	-6	-4	-2

A. $2x + y = -12$

B. $2x = -6$

C. $2x - 3y = 12$

D. $2x + 3y = 12$

E. $-3x + y = -8$

Strategy for Solving Word Problems

1. Read the problem carefully several times until you can state in your own words what is given and what the problem is looking for (decide what you are asked to find). Let x or A (or any variable) represent one of the unknown quantities in the problem.
2. If necessary, write expressions for any other unknown quantities in the problem in terms of x.
3. Translate the problem into an equation. Write an equation in terms of x that translates, or models, the conditions of the problem.
4. Solve the equation and answer the problem's question.
5. Check the solution in the original wording of the problem, not in the equation obtained from the words.

Note: Review Chart 1, 2 and 3 (Section 2.1).

Example 1) The **product** of 5 and **a number is** 120. Find the number.

1. Let x represent one of the quantities (a number).

   x = the number

2. Represent other quantities in terms of x. There are no other unknown quantities.

3. Write an equation in x that describes the conditions.

   $5*x = 120$

4. Solve the equation and answer the question.

   $5x = 120$

   $\dfrac{5x}{5} = \dfrac{120}{5}$ (Here we divided both sides by 5 so that we can find the value of x).

   $x = 24$

The number is 24.

5. Check the solution.

   The product of 5 and 24 is 120.

Example 2) What is 72% of 800?

Solution:

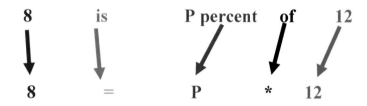

What is 72% of 800?

x = 0.72 * 800

x = 576

Example 3) 8 is what percent of 12?

Solution:

8	is	P percent	of	12
8	=	P	*	12

$8 = P * 12$

$$\frac{8}{12} = \frac{12P}{12}$$

$0.6\overline{6} = P$

$P = 67\%$ Rounded to the nearest percent.

Example 4) 57% of what number is 163.191?

Solution:

57% of what number is 163.191?

0.57 * x = 163.191

$$\frac{0.57x}{0.57} = \frac{163.191}{0.57}$$

x = 286.3

Example 5) How many tiles of 15 cm^2 will be needed to cover a floor of dimension 6.9 m length by 5.8 m width?

Solution:

    1) First convert the dimensions into cm.

        1m = 1*100 = 100 cm

Length = 6.9 m = 6.9 * 100 = 690 cm.

Width =5.8 m = 5.8 * 100 = 580 cm.

    2) Find the area of rectangle.

    Area = length * width = 690 * 580 = 400,200 cm^2

    3) Divide the area by 15.

$\dfrac{400,200}{15} = 26,680$ tiles.

Example 6) Renting a car for one day costs $40 plus $0.30 per mile. How much would it cost to rent the car for one day if 70 miles are driven?

Solution:

$40 = fixed cost,

$0.30 * 70 = variable cost

$40 + 70 × $0.30 =

$40 + $21 =

$61

Example 7) How long does a 220-mile trip take moving at 50 miles per hour (mph)?

Solution:

220m / 50 mph = 4.4 hours.

Example 8) Sean has one million dollars in his bank account. He cashes out only in $100 bills. How many $100 currency notes should he collect?

Solution:

1000, 000/100 = 10,000.

Example 9) The following graph shows the sales figures for a toy company since it opened 10 years ago. Per the graph, what was the dollar value of sales in the company's 10th year of business?

Solution:

0.8 * 1,000,000 = $800,000.

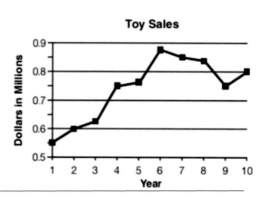

Example 10) You roll two dice 36 times. What is the most likely number of times that you will roll a 4?

Solution:

There are 6 x 6 or 36 possible combinations of outcomes for rolling a dice. Three of these outcomes result in 4:

➢ 1, 3
➢ 3, 1 and
➢ 2, 2

Therefore, the odds favor rolling four 3 times in 36 rolls.

Note) To change the decimal numbers to percentage do the following:

Move the decimal point two places to the right then put % symbol next to the number for example:

$$0.035 = 3.5\%$$

Example 11) Change each of the following decimal numbers to percentage number.

      1) 0.35   2) 0.45   3) 7.86

Solution:

1) $0.35 = 35\%$

2) $0.45 = \ 45\%$

3) $7.86 = \ \ 786\%$

Note) To change the fraction number to percentage do the following:

1) Divide the numerator by the denominator.

2) Move the decimal point two places to the right then put % symbol next to the number for example:

$$\frac{5}{8} = 0.625 = 62.5\%$$

Note) To change the percentage number to decimal do the following:

Divide that number by 100 as in the following examples:

  a) $55\% = \dfrac{55}{100} = 0.55$

  b) $87.5\% = \dfrac{87.5}{100} = 0.875$

Example 12) Express 63% as a decimal.

Solution:

$63\% = \dfrac{63}{100} = 0.63$

Example 13) What percent of 24 is 6?

$P * 24 = 6$

$24P = 6$

$$P = \frac{6}{24} = \frac{1}{4} = 0.25 = 25\%$$

The Volume of a Cylinder (V)

$$V = \pi * r^2 * h$$

Where $\pi \approx 3.14$

r is the radius,

h is the height

Example 14) Find the volume of the following cylinder that has a height of 0.090" with the radius of 0.095"

r = 0.095"

h = 0.090"

$$V = \pi * r^2 * h$$
$$= 3.14 * (0.095)^2 * 0.090"$$
$$= 0.028 * 0.090"$$
$$= 0.0025 \ in^3$$

Note:

Below, the charts are some basic mathematical formulas you may need it to solve some word problems:

Chart #1 (Volume of some mathematical shapes)

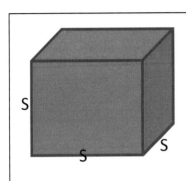

Cube

$V = S^3$

Where S is a side.

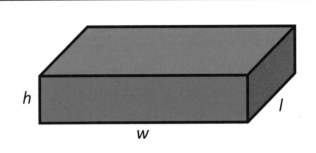

Rectangular Solid

$V = l * w. h$

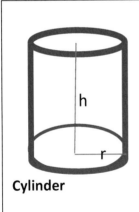

Cylinder

$V = \pi * r^2 * h$

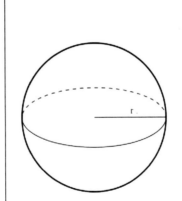

Sphere

$V = \frac{4}{3} * \pi * r^3$

Chart #2 (Area of Some Mathematical Shapes)

Area of a square

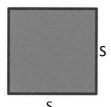

S

S

$$A = S^2$$

Where S is a side

Area of a rectangle

w

l

$$A = l * w$$

Area of a triangle

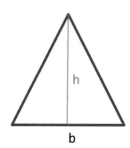

h

b

$$A = \frac{1}{2} * b * h$$

Area of parallelogram

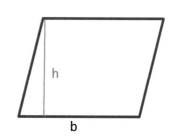

h

b

$$A = b * h$$

Area of a circle

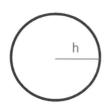

h

$$A = \pi * r^2$$

Area of a trapezoid

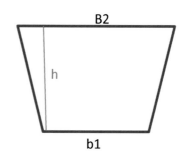

B2

h

b1

$$A = \frac{1}{2}(b1 + b2) * h$$

Chart# 3 (Perimeter of Some Mathematical Shapes)

Perimeter of a square

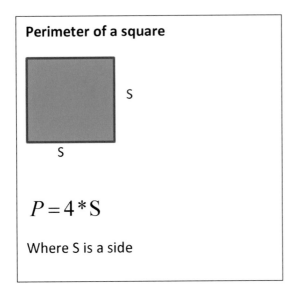

$P = 4 * S$

Where S is a side

Perimeter of a square

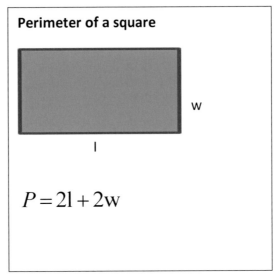

$P = 2l + 2w$

Perimeter of a triangle

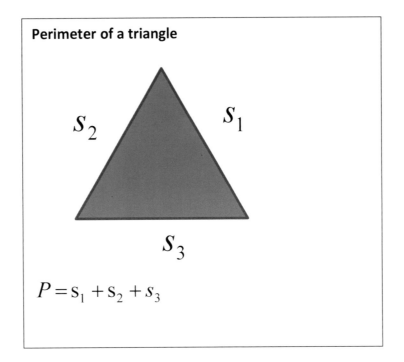

$P = s_1 + s_2 + s_3$

Example 15) what is algebraic expression that shows the average melting points of helium, hydrogen, and neon if h represents the melting point of helium, j represents the melting point of hydrogen, and k represents the melting point of neon.

<u>Solution:</u>

(h + j + k)/3.

Q1) An office uses paper drinking cups in the shape of a cone with the dimensions as shown. Find the volume of the cone.

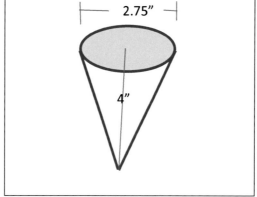

Q2) A payroll clerk is calculating employee pay for the last week based on the information shown in the table. What is the total amount of pay the employees received for the last week?

Number of Employees	Number of Hours Worked	Hourly Pay
3	40	$45
5	40	$20
2	30	$15
2	20	$12

Q3) Janice is the fund-raising manager for a local charity. She is ordering caps for an upcoming charity walk. The company that makes the caps charges $6 per cap plus a $25 shipping feed. Janice has a budget of $1,000. What is the greatest number of caps she can buy?

    A) 162   **B)** 163   **C)** 166   **D)** 167

Q4) Scott entered a cooking competition. Competitors' final scores were calculated by using a weighted average of 75% for taste and 25% for presentation. Louis received scores of 84 for taste and 68 for presentation. What was his final score?

Q5) Ms. Carla is having wood flooring putting in her rectangular bedroom. The area of the room is 234 square feet. The length of the room is 18 feet. What is the width, in feet of the room?

Q6) The graph shows the height of a ball in relation to the number of seconds since the ball was thrown. Which point represents the height of the ball at the time that it was thrown?

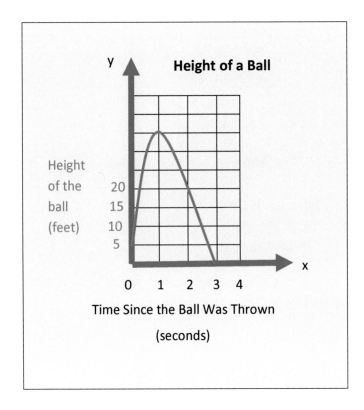

Q7) Three eighths of a garden is dug in 9 hours. How long will it take to dig to five eighths of it?

Q8) A box has a length of 30 cm. It has a width that is one third of the length. The height of box is one third of its width. What is approximate value of volume of water required to fill this box?

Q9) 19 is 50% of what number?

Q10) A bag has 7 balls: green, black, blue, brown, pink, red, and orange. If 6 balls are removed from the bag at random, what is the probability that a blue ball has been removed?

Linear Inequalities in One Variable: An inequality in one variable (usually x) that can be written in the form $ax + b < 0$ or $ax + b > 0$, where "a" and "b" are real numbers with $a \neq 0$.

The inequality symbols are:

Symbol	Read as
<	Less than
>	Greater than
≤	Less than or equal
≥	Greater or equal

Properties of Inequalities

1) The Addition Property of Inequality:

If $a < b$, then $a + c < b + c$ and

if $a < b$, then $a - c < b - c$

Important Vocabulary

Solution of an inequality: A value of the variable for which the inequality true.

Graph of an inequality: The set of all points on the real number line that represent the solution set of an inequality.

Double inequality: An inequality that represents two inequalities.

This means, if the same quantity is added to or subtracted from both sides of an inequality, the resulting inequality is equivalent to the original one. As in the following examples:

$2x + 3 < 7$

____-3__-3_   Here we subtract 3 from both sides of the above inequality.

$2x < 4.$

And

$5x - 1 > 2$

___+1__+1   Here we add 1 to both sides of the above inequality.

$5x > 3$

2) The Positive Multiplication Property of Inequality:

If a < b and c is any positive number, then

a * c < b * c

For example:

If x < 5 then

x * 3 < 5 * 3

3) The Positive Division Property of Inequality:

If a < b and c is any positive number, then

a ÷ c < b ÷ c   or we can write it as:

$$\frac{a}{c} < \frac{b}{c}$$

For example:

x < 3 then

$$\frac{x}{2} < \frac{3}{2}$$

> **Note:**
>
> The sense of the inequality is not changed if both sides are multiplied or divided by the same positive number.

Note: If we multiply or divide both sides of an inequality by the same positive quantity, the resulting inequality is equivalent to the original one. As in the following examples:

2x < 5

$$\frac{2x}{2} < \frac{5}{2}$$ Here we divide both sides by positive 2.

$$\frac{x}{3} < 9$$

$$3 * \frac{x}{3} < 3 * 9$$ Here we multiply both sides by positive 3.

4) The Negative Multiplication Property of Inequality:

If a < b and c is any negative number, then

a * c > b * c

For example:

if x < 5 then

x * –3 > 5 * –3

5) The negative Division Property of Inequality:

If a < b and c is any negative number, then

a ÷ c > b ÷ c   or we can write it as: $\dfrac{a}{c} > \dfrac{b}{c}$

For example:

x < 3 then $\dfrac{x}{-2} > \dfrac{3}{-2}$

> **Important Note:** If we multiply or divide both sides of an inequality by the same negative quantity, you must reverse the direction of the inequality symbol.

For example:

-4x < 5

$\dfrac{-4x}{-4} > \dfrac{5}{-4}$ Here we divide both sides by -4 that is why we reverse the inequality from less than to greater than.

Flipping Rule:

Anytime an inequality is multiplied or divided by a negative number the inequality must be reversed.

For example: $\dfrac{x}{-2} > 9$

$-2 * \dfrac{x}{-2} < -2 * 9$ Here we multiplied both sides by -2 that is why we reverse the inequality symbol.

Graphs of Inequalities

1) The graph of inequality containing the inequality symbol ">" is open parenthesis symbol "(" for example: The graph of inequality $x > 3$ will be:

$$3$$

You can also graph it as in the following:

$$3$$

The red arrow shows that the graph extends indefinitely to the right.

This is an open circle because the mathematical statement is $>$ rather than $\geq$.

2) The graph of inequality containing the inequality symbol "<" is close parenthesis symbol ")" for example: The graph of inequality $x < 0$ will be:

$$0$$

You can also graph it as in the following:

$$0$$

The red arrow shows that the graph extends indefinitely to the left.

This is an open circle because the mathematical statement is $<$ rather than $\leq$.

3) The graph of inequality containing the inequality symbol "≥" is open bracket symbol "[" for example: The graph of inequality x ≥ 4 will be:

4

You can also graph it as in the following:

4

The red arrow shows that the graph extends indefinitely to the right.

This is a closed circle because the mathematical statement says ≥ rather than >.

4) The graph of inequality containing the inequality symbol "≤" is close bracket symbol "]" for example: The graph of inequality x ≤ 2 will be:

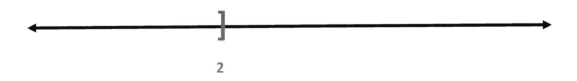

2

You can also graph it as in the following:

2

This is a closed circle because the mathematical statement says ≤ rather than <.

The following table shows four inequalities, their solution sets using interval and set builder notations, and graphs of the solution sets:

Inequality	Interval Notation	Set-Builder Notation	Graph
X > a	(a, ∞)	{x l x > a}	a ◄———(——►
X ≥ a	[a, ∞)	{x l x ≥ a}	a ◄———[——►
X < b	(-∞, b)	{x l x < b}	b ◄———)——►
X ≥ b	(-∞, b]	{x l x ≤ b}	b ◄———]——►

Strategy for Solving Linear Inequalities in one Variable

1. Multiply to clear fractions by using the multiplication properties of inequality.

2. Simplify and use distributive property, if needed. This will involve things like removing parentheses.

3. Get all variable terms on one side and numbers on the other side of inequality by using the addition property of inequality.

4. Combine the like terms, if needed.

5. Isolate variable by using the multiplication properties of inequality.

Example 1) Solve for x in the inequality 2x +3 < 9 then graph it.

Solution:

2x + 3 < 9
    -3   -3 Here we subtract 3 from both sides.

2x < 6

$\dfrac{2x}{2} < \dfrac{6}{2}$ Here we divide both sides by 2.

x < 3

The solution is x < 3 states that any number less than 3 makes the inequality true.
The following is the graph:

3

Example 2) Solve for x in the inequality 2x –5 > 7 then graph it.

Solution:

2x – 5 > 7
   +5   +5 Here we add 5 for both sides.

2x > 12

$\dfrac{2x}{2} > \dfrac{12}{2}$ Here we divide both sides by 2.

x > 6

The solution is x > 6 states that any number greater than 6 makes the inequality true.
The following is the graph:

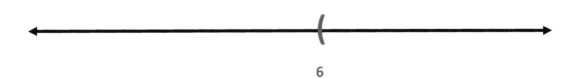

6

Example 3) Solve for x in the inequality -4x +3 ≤ 23 then graph it.

<u>Solution:</u>

-4x + 3 ≤ 23

   -3   -3   Here we subtract 3 from both sides.

 -4x ≤ 20

$\dfrac{-4x}{-4} \geq \dfrac{20}{-4}$ Here we divide both sides by -4 and we flip the inequality symbol.

x ≥ -5

The solution is x ≥ -5 states that any number less than 3 makes the inequality true. The following is the graph:

-5

Compound Inequality: A compound inequality is two inequalities joined together. For example:

$$0 \leq 2(5 - x) < 16$$

To solve the compound inequality, perform operations simultaneously to all three parts of the inequality (left, middle and right).

Example 4) Solve for x in the following compound inequality then graph it.

$0 \leq 2(5 - x) < 12$

<u>Solution:</u>

$0 < 2(5 - x) \leq 12$

$0 < 10 - 2x \leq 12$ Here we used the distributive property.

$-10 + 0 < -10 + 10 - 2x) \leq 12 - 10$ Here we subtract 10 from each side.

$-10 < -2x \leq 2$ Here we combine the like terms (simplified).

   $5 > x \geq -1$ Here we divide each part by –2 and we flip the inequality symbols because we divide by a negative

number (-2). Below is the graph for the above compound inequality:

-1   5

Q1) Solve for x in the inequality $x + 7 \geq 9$.

Q2) Solve the following for h: $-h - 11 > 23$.

Q3) Solve for y in the inequality $-3y + 5 > 23$.

Q4) Solve the following compound inequality then graph it:

$0 \leq 4(5 - x) < 8$.

Q5) Daniel volunteers and wants to average between 90 and 95 hours per month. The numbers of hours that he volunteered for the last three months, are 84, 94, and 91. Daniel solves the inequality

$$90 \leq \frac{84 + 94 + 91 + h}{4} \leq 95$$

To determine the number of hours, he needs to volunteer during the fourth month to reach his desired average. What is the solution to the inequality?

    A. $90 \leq h \leq 95$

    B. $90 \leq h \leq 96$

    C. $92 \leq h \leq 94$

    D. $91 \leq h \leq 111$

Q6) The store employee works 35 hours per week. Which inequality can be used to find the dollar value, x of weekly sales that the employee must make to earn more than $400 per week?

    A. $35(8) + 0.08x < 400$

    B. $35(8) + 0.08x > 400$

    C. $35(0.08) + 8x < 400$

    D. $35(0.08) + 8x > 400$

Q7) Solve the following inequality:

$\dfrac{-1}{7}x \geq 0$.

Chapter 3

Exponents and Rational Expressions

Section 3.1 Exponents

When a number, variable, or expression is raised to a power, number, variable, or expression is called the base and the power is called the exponent. As in the following example:

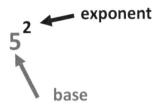

Here the exponent form is 5^2.

The factor form for 5^2 is 5 * 5.

Exponents Vocabulary

Factors are numbers that are multiplied together. So, 5 * 5 are factors of 5^2. Base is the number being used as a factor. 5 is the base of 5^2.

Exponent tells how many times a number is used as a factor. 2 is the exponent (sometimes we call it an index, or power) of 5^2. Below table contains some examples of Exponent Form, Factor Form, and Standard Form:

Exponent Form	Factor Form	Standard Form
5^3	$5*5*5$	125
x^2	$x*x$	x^2
2^5	$2*2*2*2*2$	32

Exponents' Properties

1) If **a** is any real number and **r** and **s** are integers, then

$$a^r * a^s = a^{r+s}$$

To multiply with the same base, add exponents and use the common base.

Example 1) Multiply $4^2 * 4^1$.

<u>Solution:</u>

$4^2 * 4^1 = 4^{2+1} = 4^3 = 4*4*4 = 64$.

Notice that $4^2 = 16$ and $4^1 = 4$ and $16*4 = 64$.

2) If **a** is any real number and **r** and **s** are integers, then

$$\left(a^r\right)^s = a^{r \cdot s}$$

A power raised to another power is the base raised to the product of the powers.

Example 2) Multiply $(5^2)^3$

<u>Solution:</u>

$(5^2)^3 = 5^{2*3} = 5^6 = 15,625$.

3) If *a* and *b* are any real numbers and *r* is an integer, then

$$\left(ab\right)^r = a^r \cdot b^r$$

Example 3) $(5x)^2 = ?$
<u>Solution:</u>

$$\left(5x\right)^2 = 5^2 x^2 = 25x^2.$$

4) If **a** is any real number and **r** and **s** are integers, then

$$\frac{a^r}{a^s} = a^{r-s} \quad (a \neq 0)$$

To divide with the same base, subtract the exponents and use the common base.

Example 4) Simplify

$$\frac{a^3}{a^2}.$$

Solution:

$$\frac{a^3}{a^2} = a^{3-2} = a^1 = a.$$

Definition: If **r** is a positive integer number, then

$$a^{-r} = \frac{1}{a^r} \quad (a \neq 0).$$

A negative exponent in the numerator requires moving the base to the denominator, and vice-versa.

Example 5) Simplify a^{-3}.

Solution:

$$a^{-3} = \frac{1}{a^3}.$$

Example 6) Simplify $3x^{-6}$.

Solution:

$$3x^{-6} = \frac{3}{x^6}.$$

Notice that exponent does not touch number 3.

Notes
❑ Positive exponents indicate numbers greater than 1.
$7^2 = 7*7 = 49$
❑ Negative exponents indicate numbers less than 1.
$2^{-3} = \frac{1}{2^3} = \frac{1}{8} = 0.125$

❑ When a negative number appears as an exponent, switch the position of the base (from numerator to denominator or denominator to numerator) and make the exponent positive. *The sign of the base does not change.*

5) If **a** and **b** are any two real numbers and **r** is an integer, then

$$\left(\frac{a}{b}\right)^r = \frac{a^r}{b^r}$$

Distribute the exponent.

Example 7) Simplify

$$\left(\frac{5}{y}\right)^2.$$

Solution:

$$\left(\frac{5}{y}\right)^2 = \frac{5^2}{y^2} = \frac{25}{y^2}.$$

Zero Law of Exponents: Any nonzero number or variable raised to the ZERO Power = ONE. As in the following examples:

$x^0 = 1$

$9588^0 = 1$

$(-5)^0 = 1$

$(\frac{3}{4})^0 = 1$

$(2xy^2)^0 = 1.$

Example 8) Simplify $4^0 + 4^1$.

Solution:

$4^0 + 4^1 = 1 + 4 = 5.$

Notes

❑ Any number raised to the zero power is 1:

$$10^0 = 1$$

❑ Any number raised to the power of 1 equals itself:

$$10^1 = 10$$

Example 9) The Milky Way Galaxy is about 100,000 light years wide (A light year is the distance light travels in one year). Write 100,000 as an exponent form with 10 as the base.

Solution:

$10^5.$

Exponents Summary Review

$$a^r \cdot a^s = a^{r+s}$$

$$\left(a^r\right)^s = a^{r \cdot s}$$

$$\frac{a^r}{a^s} = a^{r-s}$$

$$(ab)^r = a^r b^r$$

$$\left(\frac{a}{b}\right)^r = \frac{a^r}{b^r}$$

$$a^{-r} = \frac{1}{a^r} \qquad a \neq 0$$

$$a^1 = a$$

$$a^0 = 1 \qquad a \neq 0$$

Q1) Write each of the following exponent forms in standard:

    a. 10^0

    b. 10^1

    c. 10^2

    d. 10^3

Q2) Multiply $x^2 * x^8$.

Q3) Multiply $(x^2)^5$.

Q4) $(4a)^2 = ?$

Q5) Simplify $\dfrac{x^{18}}{x^{15}}$.

Q6) Divide $\dfrac{18\,s^{12} * t^5}{36\,s^{11} * t^2}$.

Q7) $x^{-1} = ?$

Q8) Simplify $\dfrac{(x^3 x^5)^2}{x^9}$.

1) Adding and Subtracting Rational Expressions with the same Denominator

Rational expressions can be written in the form $\dfrac{P}{Q}$ where P and Q are both polynomials and $Q \neq 0$. The followings are examples of rational expressions:

$$\frac{1}{x}, \frac{3x^2 - 2x + 12}{5x + 9} \text{ and } \frac{8x^2}{5}.$$

Adding Rational Expressions

If $\dfrac{P}{Q}$ and $\dfrac{R}{Q}$ are rational expressions, then

$$\boxed{\frac{P}{Q}} + \boxed{\frac{R}{Q}} = \boxed{\frac{P+R}{Q}}$$

Note: To add the numerators, the denominators must be the same.

Example 1) Simplify the following:

$$\frac{7}{x} + \frac{2}{x}.$$

<u>Solution:</u>

$$\frac{7}{x} + \frac{2}{x} = \frac{7+2}{x} = \frac{9}{x}.$$

Subtracting Rational Expressions

If $\dfrac{P}{Q}$ and $\dfrac{R}{Q}$ are rational expressions, then

$$\dfrac{P}{Q} - \dfrac{R}{Q} = \dfrac{P-R}{Q}$$

Note: To subtract the numerators, the denominators must be the same.

Example 2) Simplify the following:

$$\dfrac{x}{y+1} - \dfrac{9}{y+1}.$$

Solution:

$$\dfrac{x}{y+1} - \dfrac{9}{y+1} = \dfrac{x-9}{y+1}.$$

Evaluating Rational Expressions

To evaluate a rational expression for a value (or values), substitute the replacement value(s) into the rational expression and simplify the result.

Example 3) Evaluate the following expression for $x = 5$:

$$\dfrac{3}{2x}.$$

Solution:

$$\dfrac{3}{2x} = \dfrac{3}{2(5)} = \dfrac{3}{10} \quad \text{Here we replace x with 5.}$$

Example 4) Evaluate the following expression for $y = 3$:

$$\dfrac{y^2}{2}.$$

Solution:

$$\dfrac{y^2}{2} = \dfrac{(3)^2}{2} = \dfrac{9}{2} \quad \text{Here we replace y with 3.}$$

2) Adding and Subtracting Rational Expressions with the Unlike Denominators

To add or subtract rational expressions with unlike denominators, you should change them to equivalent forms that have the same denominator (a common denominator). This involves finding the least common denominator of the two original rational expressions.

Least Common Denominator (LCD): The least common denominator (LCD) for a set of denominators is the smallest quantity that is exactly divisible by each denominator.

To find a Least Common Denominator of two rational expressions, you should do the following:

1) Factor the given denominators.

2) Take the product of all the unique factors.

Each factor should be raised to a power equal to the greatest number of times that factor appears in any one of the factored denominators.

Example 5) Find the LCD between

$$(x-2)^2 \text{ and } (x-2).$$

Solution:

$$LCD = (x-2)^2.$$

Example 6) Find the LCD between

$$(x-3) \text{ and } (x+2).$$

Solution:

$$LCD = (x-3)*(x+2).$$

Example 7) Find the LCD between

$2x$ and x^5.

Solution:

$LCD = 2x^5.$

Steps to Add or Subtract Fractions:

Step 1: Factor each denominator completely and use the factors to build the LCD.

Step 2: Rewrite each fraction as an equivalent fraction that has the LCD for the denominator.

Step 3: Add or subtract the numerators produced in step 3.

Step 4: Reduce and simplify the fraction if possible.

Example 8) Simplify $\dfrac{3}{4x} + \dfrac{5}{2x^2}$.

<u>Solution:</u>

    1) Factor each denominator completely then finds the LCD between both denominators 4x and $2x^2$

$4x = 2*2*x = 2^2 x$

$4x^2 = 2*2*x^2 = 2^2 x^2$

$LCD = 2^2 x^2 = 4x^2$.

    2) Rewrite each fraction as an equivalent fraction that has the LCD for the denominator.

$\qquad \dfrac{3}{4x} = \dfrac{3}{4x} * \dfrac{x}{x} = \dfrac{3x}{4x^2}$ Here the missing factor is x.

$\qquad \dfrac{5}{2x^2} = \dfrac{5}{2x^2} * \dfrac{2}{2} = \dfrac{10}{4x^2}$ Here the missing factor is 2.

    3) Put the same denominator then add numerators.

$\dfrac{3}{4x} + \dfrac{5}{2x^2} = \dfrac{3x}{4x^2} + \dfrac{10}{4x^2} = \dfrac{3x+10}{4x^2}$.

Q1) Simplify the following:

$$\frac{20}{(x-2)} - \frac{19}{(x-2)}.$$

Q2) Simplify

$$\frac{6}{5x} + \frac{7}{10x^2}.$$

Q3) Evaluate the following expression for $x = 0$:

$$\frac{10}{(x+2)}.$$

Q4) Evaluate the following expression for $x = -8$:

$$\frac{20}{(x-2)}.$$

Q5) Find the LCD between $(x-2)(x-3)$ and $(x-2)^2$.

Q6) Simplify the following rational expression:

$$\frac{3}{x+2} - \frac{8}{x-2}.$$

Q7) Add the following rational expressions:

$$\frac{6}{5x} + \frac{7}{10x^2}.$$

Rules for Rational Expressions (Fractions)

1) For any algebraic expressions A, B, X, and Y. B and X do not equal zero:

$$\frac{A}{B} * \frac{Y}{X} = \frac{A*Y}{B*X}$$

Example 1) Multiply the following:

$$\frac{2}{3} * \frac{x}{y}.$$

<u>Solution:</u>

$$\frac{2}{3} * \frac{x}{y} = \frac{2*x}{3*y} = \frac{2x}{3y}.$$

> **Note:**
>
> To Multiply a rational expression:
>
> **1.** Factor all numerators and denominators.
>
> **2.** Cancel all common factors.
>
> **3.** Either multiply the denominators and numerators together or leave the solution in factored form.

Definition:

Two numbers whose product is 1 are called reciprocals as in the following examples:

The reciprocal of $\frac{a}{b}$ is $\frac{b}{a}$,

the reciprocal of $\frac{7b^5}{8r^2a}$ is $\frac{8r^2a}{7b^5}$, and

the reciprocal of $\frac{3xy^2t^4}{2a^5b^7}$ is $\frac{2a^5b^7}{3xy^2t^4}$.

> **Note:**
>
> To get the reciprocal of a fraction, just turn it upside down. In other words, swap over the Numerator and Denominator.

Fundamental Property of Rational Expressions

If P/Q is a rational expression and if k represents any factor where $k \neq 0$, then

$$\frac{P\cancel{k}}{Q\cancel{k}} = \frac{P}{Q}.$$ In other words, this will let us reduce fractions.

2) For any algebraic expressions A, B, X, and Y. B, X, and Y do not equal zero:

$$\frac{A}{B} \div \frac{X}{Y} = \frac{A}{B} * \frac{Y}{X} = \frac{A*Y}{B*X}.$$

In other words: Multiply the first rational expression by the reciprocal of the second rational expression. Or you can use the KCF method where;

K for Keep the first fraction.
C for Change the division to multiplication.
F for Flip the 2nd fraction.

Example 2) Divide the following: $-\frac{2}{x} \div \frac{5}{x}$.

Negative Signs

$$-\frac{a}{b} = \frac{-a}{b} = \frac{a}{-b}$$

$$-\frac{2}{x} = \frac{-2}{x} = \frac{2}{-x}$$

<u>Solution:</u>

$$-\frac{2}{x} \div \frac{5}{x} = \frac{-2}{x} * \frac{x}{5} = \frac{-2}{5}.$$ Here we use the KCF method.

Excluding Values from Rational Expressions

If a variable in a rational expression is replaced by a number that makes the denominator 0, that number is excluded as a replacement for the variable. The rational expression is undefined at any value that produces a denominator of 0.

Reminder

A rational expression is undefined for all values that make the denominator 0.

Example 3) What is the excluded value for the following rational expression:

$$\frac{1}{2x - 5}.$$

<u>Solution:</u>

$2x - 5 = 0$. Here we set the denominator equal to 0.

$+5 \quad +5$. Here we add +5 to both sides.

$2x = 5$.

$\frac{2x}{2} = \frac{5}{2}$. Here we divide both sides by 2.

$x = \frac{5}{2} = 2.5$. The expression is undefined for x = 2.5.

Note:

Excluded values are values that will make the denominator of a fraction equal to 0. You can't divide by 0, so it's very important to find these excluded values when you're solving a rational expression.

To find the excluded value for a rational expression, do the following:

1) Set the denominator equal to 0.
2) Solve the resulting equation.

Example 4) State the restriction of the following rational expression:

$$\frac{9}{x}.$$

<u>Solution:</u>

x = 0.

Zero-Factor Property

If *a* and *b* are real numbers and if *ab* = 0, then *a* = 0 or *b* = 0.

Example 5) Solve the following equation using the Zero-Factor Property:

(x - 5) (x + 6) = 0.

<u>Solution:</u>

(x - 5) (x + 6) = 0

x – 5 = 0 or x +6 = 0

x = 5 or x = -6.

Note:
The zero-product property let us split the product of factors into separate equations. Then, we can solve each equation to get the solutions to our original equation!

Example 6) The volume of a packing box is 5 – x cubic feet. The width of the box is x feet and the length is x – 7 feet. The height is given by expression $\frac{5-x}{x(x-7)}$. What are the values of x that make the expression undefined?

<u>Solution:</u>

x (x–7) = 0

x = 0 or x –7 = 0

x = 0 or x = 7

x = {0, 7}.

Q1) Multiply the following:

$$\frac{3x}{y^2} * \frac{4x}{y}.$$

Q2) Divide the following:

$$\frac{8x^5}{3} \div \frac{2x^2}{9}.$$

Q3) What are the excluded values for the following rational expression?

$$\frac{5x^2 + 4x - 13}{(x-4)(x+7)}.$$

Q4) For what value of x is the following expression undefined?

$$y = \frac{45x^2 + 23x - 49}{15 - x}.$$

Q5) Divide the following:

$$\frac{16x^2}{10x^3} \div \frac{12}{5x}.$$

Q6) Solve the following equation for x:

$$3ax = \frac{1}{5}.$$

Q7) Divide the following:

$$\frac{8x5}{3} \div \frac{2x^2}{9}.$$

Q8) Solve the following equation:

$$(3x - 2)(4x + 8) = 0.$$

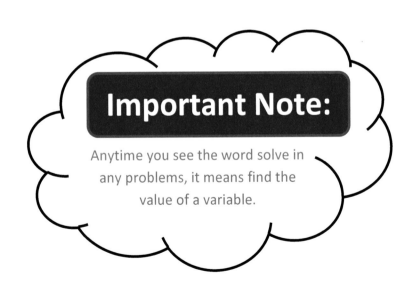

Important Note:

Anytime you see the word solve in any problems, it means find the value of a variable.

Evaluate Rational Expressions

To evaluate a rational expression for a value (or values), substitute the replacement value(s) into the rational expression and simplify the result.

Example 1) Evaluate the following expression for $x = 2$:

$$\frac{x+10}{4}.$$

Here we replace x with 2.

Solution:

$$\frac{x+10}{4} = \frac{2+10}{4} = \frac{12}{4} = 3.$$

Example 2) Evaluate the following expression for $x = 5$:

$$\frac{2(x)+20}{3}.$$

Solution:

$$\frac{2(x)+20}{3} = \frac{2(5)+20}{3} = \frac{10+20}{3} = \frac{30}{3} = 10.$$

Example 3) Evaluate the following expression for $x = 4$:

$x^2 - 2(x) + 3$.

Solution:

$x^2 - 2(x) + 3 = (4)^2 - 2(4) + 3 = 16 - 8 + 3 = 11.$

Here we replace x with 4.

Simplify Rational Expressions

To simplify rational expressions, do the following steps:

1) Factor the numerator and denominator completely.
2) Divide both the numerator and denominator by any common factors.

Example 4) Simplify the following:

$$\frac{21x^2}{15x}.$$

Here the factor of 21 is 3*7, the factor of $x^2 = x * x$, and the factor of 15x is 3*5*x.

Solution:

$$\frac{21x^2}{15x} = \frac{3 * 7 * x * x}{3 * 5 * x} = \frac{7x}{5}.$$

Example 5) Simplify the following:

$$\frac{25x^3}{15x} * \frac{3xy}{y^2}.$$

Solution:

$$\frac{25x^3}{15x} * \frac{3xy}{y^2} = \frac{5 * 5 * x * x * x * 3 * x * y}{3 * 5 * x * y * y} = \frac{5x^3}{y}.$$

Exercise Set 3.4

Q1) Evaluate the following expression for x = 5:

$x^3 - 124$.

Q2) $\dfrac{(x-2)^3}{5x} = ?$ *when* x = 4.

Q3) Simplify the following rational expression $\dfrac{4A^2}{AB^3} * \dfrac{3BC}{12A^3}$.

Q4) Simplify $\dfrac{4x^2}{2x} * \dfrac{3y}{6xy}$.

Chapter 4

Polynomials and Factoring

Section 4.1 Adding and Subtracting Polynomials

Definitions:

A **monomial** is a real number, variable, or product of real numbers, and one or more variables. The followings are examples of monomials:

- ❖ -3
- ❖ 15x
- ❖ $-23x^2y$
- ❖ $\frac{3}{4}a^2b^5$

A polynomial with exactly two terms (two monomials is called a **binomial**. The followings are examples of binomials:

- ❖ x + y
- ❖ 15x – 3y
- ❖ $24x^2y + 15$
- ❖ $\frac{3}{4}a^2b^5 + 24x^2y$

> A **binomial** is a sum of **two** monomials.

A polynomial with exactly three terms is called a **trinomial**. The followings are examples of trinomials:

- ❖ $x^2 + 4x - 5$
- ❖ $4x^2 - 3x + 1$
- ❖ $4y^2 - 3y + 1$
- ❖ $4y^2 - 3xy + y^3$

> A **trinomial** is a sum of **three** monomials.

> A polynomial is a sum of monomials.

Finding the degree of a polynomial.

1) The degree of a term with one variable is the highest exponent on the variable. The greatest degree of any term in a polynomial is called the "degree of the polynomial". The degree of the following polynomial $x^3 + 4x - 6$ is 3 because the highest exponent on the variable x is 3.

The greatest degree of $-7x^4 - 6x + 2x^3 + x^5$ is 5 since x^5 has the highest degree of all terms.

2) The degree of a term with more than one variable is defined to be the sum of the exponent on the variables. The degree of the following monomial:

$-7t^2s^3$

is 5 because the sum of the exponents of variable **t** and **s** is 2 + 3 = 5.

Below are some examples of a standard form polynomial, binomial, and trinomials with their degrees:

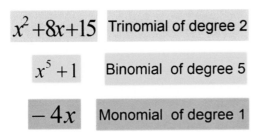

$x^2 + 8x + 15$ Trinomial of degree 2

$x^5 + 1$ Binomial of degree 5

$-4x$ Monomial of degree 1

Example 1) Find the degree of the following:

A) $-3x^4 - 6x + 2x^3$

B) $3x^4 - 6x + 2x^3 + x^5$

C) $6x$

D) -2

E) $4m^3 - 32m^3p^6$

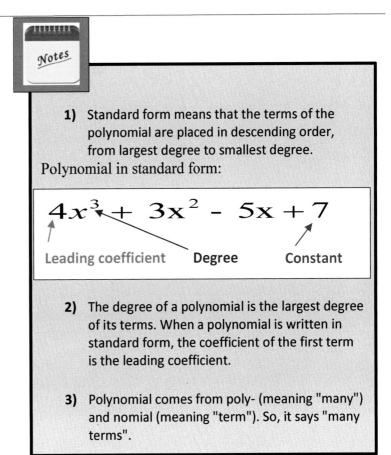

1) Standard form means that the terms of the polynomial are placed in descending order, from largest degree to smallest degree.

Polynomial in standard form:

$$4x^3 + 3x^2 - 5x + 7$$

Leading coefficient   Degree   Constant

2) The degree of a polynomial is the largest degree of its terms. When a polynomial is written in standard form, the coefficient of the first term is the leading coefficient.

3) Polynomial comes from poly- (meaning "many") and nomial (meaning "term"). So, it says "many terms".

Solution:

A) $-3x^4 - 6x + 2x^3$ degree = 4

B) $3x^4 - 6x + 2x^3 + x^5$ degree = 5

C) $6x$ degree = 1

D) -2 degree = 0

E) $4m^3 - 32m^3p^6$ degree = 9

The following table contains some examples of the polynomials classification:

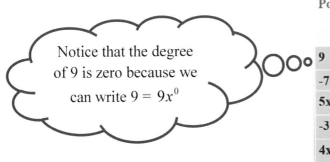

Notice that the degree of 9 is zero because we can write $9 = 9x^0$

Polynomial	Polynomial Degree	Classify Polynomial Terms
9	0	Monomial
-7x	1	Monomial
5x+22	1	Binomial
-3x²+2x-5	2	Trinomial
4x⁴-8x	4	Binomial
5x⁵-8x³-6x+1	5	Polynomial

To add or subtract polynomials:

      1) Align the like terms (place the like terms together)

      2) Add/Subtract the like terms

Note: When adding or subtracting, exponents remain the same.

Example 2) Add $4x^2 + 5x^2$.

<u>Solution</u>:

$4x^2 + 5x^2 = 9x^2$.

Example 3) Add the following polynomials then rewrite simplified polynomial so that the powers of x are in descending form.

$(2x^3 + 5x^2 + 4x^5) + (6x^3 + x^2 + 9x^5)$.

<u>Solution</u>:

$(2x^3 + 5x^2 + 4x^5) + (6x^3 + x^2 + 9x^5) = (2x^3 + 6x^3) + (5x^2 + x^2) + (4x^5 + 9x^5)$

$$= \quad 8x^3 \qquad + 6x^2 \qquad + \quad 13x^5$$

$$= 13x^5 \ + \ 8x^3 \ + 6x^2$$

Example 4) Subtract the following polynomials:

$(19a + 8b + 17c) - 2(4a - 3b + 5c)$.

> Here we multiply -2 by each term in $(4a - 3b + 5c)$
>
> so $-2(4a - 3b + 5c) = -8a + 6b - 10c$

<u>Solution</u>:

$(19a + 8b + 17c) - 2(4a - 3b + 5c) = 19a + 8b + 17c - 8a + 6b - 10c$

$$= (19a - 8a) + (8b + 6b) + (17c - 10c)$$

$$= 11a + 14b + 7c.$$

> Here we combine the like

106

SUMMARY

1) A monomial is a real number, variable, or product of real numbers, and one or more variables.

2) A binomial is a sum of two monomials.

3) A trinomial is a sum of three monomials.

4) A polynomial is a sum of monomials.

5) Degree of a polynomial: the degree of a polynomial is the largest degree of its terms.

6) Standard form: terms are written in descending order from the largest to the smallest degree.

7) Coefficient: the integer in front of the variable. How many you have of each variable? If no number, you have one.

Exercise Set 4.1

Q1) Add the following polynomials (9y − 7x + 15a) + (-3y + 8x − 8a).

Q2) Find the sum of the followin polynomials (5x − 3y) + (2x + 6y).

Q3) Subtract the following polynimials (8x + 6y) − (x − 3y).

Q4) Rewrite the following polynomial so that the powers of x are in descending form:

$-8x^2 + 3x^5 + 1 + 5x + 7x^4 + 4x^3$.

Q5) Subtrcat $(2x^2 - 5y^2 + 7x)$ from $(5x^2 + 3y^2 - 5x)$.

Q6) Subtract $(x^2 + 5y^2 - 7x - 2)$ from $(5x^2 - 12y^2 - 5x + 3)$.

Q7) Classify the following by degree and term:

$4y^2 - 3y + 1$.

Q8) Find the area of the following shaded region:

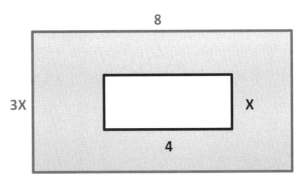

A binomial is a polynomial with two terms. To multiply a binomial by a binomial, you will use a method called "**FOIL**". This process is called "FOIL" because you work the problem in this order:

First

Outer

Inner

Last

Example 1) Find the product of the following using the "FOIL" method: $(x + 2) * (x + 3)$.

<u>Solution:</u>

➢ First= multiply the first term in each binomial = $x * x = x^2$.

➢ Outer= multiply the first term in the first binomial by the

   last term in the second binomial (the two terms that are farthest apart) = $x*3= 3x$.

➢ Inner = multiply the last term in the first binomial by the first term in the second binomial (the terms that are closest together) =$2*x = 2$ x then combine the like terms between outer and inner if possible.

➢ Last= multiply the last term in each binomial = $2 * 3 = 6.$

$$(x + 2)\ (x + 3) =$$

$$2x$$
$$3x$$

$$x^2 +\ 3x + 2x\ + 6 =$$

$$x^2 + 5x + 6$$

Example 2) Find the product of the following:

$$(3x + 2)\,(4x + 5).$$

Solution:

F = First terms = 3x *4x = $12x^2$

O = Outer terms = 3x * 5 = 15x

I = Inner terms = 2 * 4x = 8x

L = Last terms = 2 * 5 = 10

$$(3x + 2)\,(4x + 5) = 12x^2 + 15x + 8x + 10 = 12x^2 + 23x + 10$$

Here we combine the like terms. 15x+8x=23x.

Note: To multiply (+3x) * (+4x) do the following steps:

1) Multiply sign * sign which is (+) * (+) = +
2) Number * Number which is 3*4=12
3) Variable * Variable which is x*x= x^2

Example 3) Multiply (-2y) * (+5y).

Solution:

$(-2y)\,(+5y) = -10\,y^2.$

Here we

1) Multiply sign * sign which is (−) *(+) = −
2) Number * Number which is 2*5=10
3) Variable * Variable which is y*y= y^2

Example 4) Multiply (x - 2y) (x + 5y).

Solution:

(x - 2y) (x + 5y) =

$$x^2 + 5xy - 2xy - 10y^2 =$$

Here we combine the like terms 5xy–2xy = +3xy

$$x^2 \quad + 3xy \quad - 10y^2.$$

Example 5) Divide the following:

$$\frac{x}{(x-4)} \div \frac{5x}{(x-4)}.$$

<u>Solution:</u>

$$\frac{x}{(x-4)} \div \frac{5x}{(x-4)} = \frac{\cancel{x}}{\cancel{(x-4)}} * \frac{\cancel{(x-4)}}{5x} = \frac{1}{5}.$$

SUMMARY

The **FOIL** method is ONLY used when you multiply 2 binomials such as (x+2) (x+5).

F tells you to multiply the FIRST terms of each binomial ($x*x=x^2$).

O tells you to multiply the OUTER terms of each binomial ($5*x=5x$).

I tells you to multiply the INNER terms of each binomial ($2*x=2x$).

L tells you to multiply the LAST terms of each binomial ($2*5=10$).

Exercise Set 4.2

Q2) What is the product of (y + 7) (y + 2)?

Q3) What is the product of (r – 8) (r – 5)?

Q4) $(t+u)^2 = ?$

Q5) $(2m-p)^2 = ?$

Q6) Multiply $(3k - \frac{1}{2})(3k - \frac{1}{2})$.

Q7) Multiply (4x + 3) (2x + 1).

Q8) Multiply (3k - 2) (2k + 1).

Q9) Multiply (x–3) (x+3).

Q10) Divide $\frac{(x+2)}{3} \div \frac{(x+2)}{12}$.

Section 4.3 Prime Factorization and the Greatest Common Factor

Factors are the numbers you multiply together to get a product. A prime number is a whole number greater than 1 that has exactly two positive factors, 1 and itself. In other words, a prime number is a number that can only be divided (without remaining) by one and itself. 3 is a prime number because its only positive factors are 1 and 3.

A composite number is a whole number that has more than two positive factors. 6 is a composite number because it has more than two positive factors 1, 2, 3, and 6. In other words a composite number is a number that can be divided by another number without remaining in addition to 1 and itself such as number 6 that can be divided by 1, 2, 3, and 6.

A composite number can be written as the product of its prime factors. This is called the prime factorization of the number.

Example 1) Write the prime factorization of 18.

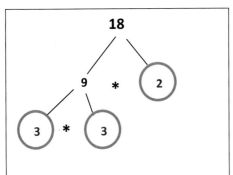

Note:

1) Write 18 as the product of two positive factors.
2) Continue factoring until all factors are prime.
3) Circle your prime numbers.
4) Write the prime factorization using exponents.

The prime factorization of 18 is 2 * 3 * 3 or $18 = 2*3*3 = 2 * 3^2$.

Example 2) Write the prime factorization of 24.

Solution:

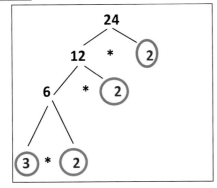

Note:

1) Write 24 as the product of two positive factors.
2) Continue factoring until all factors are prime.
3) Circle your prime numbers.
4) Write the prime factorization using exponents.

The prime factorization of 24 is 2 * 2 * 2 * 3 or $24 = 2*2*2*3 = 2^3 * 3$.

The Greatest Common Factor (GCF)

Greatest Common Factor is the largest factor that two or more numbers have in common.

Three Steps to find the GCF (Greatest Common Factor) between numbers.

Step#1 Find the factors of each number.

Step#2 Circle the common factors of the numbers.

Step#3 Multiply these common factors.

Example 3) Find the GCF of 18 and 24.

Solution:

18 = ②* 3 *③

24 = ② * 2 * 2 * ③

GCF = 2 * 3 = 6.

GCF is the Greatest Common Factor where:

G is for Greatest (largest).

C is for Common (shared).

F is for Factor (a number that divides into a whole number without remaining).

Example 4) Find the GCF of 30, 20, and 15.

Solution:

30 = 2 * 3 *⑤

20 = 2 * 2 *⑤

15 = 3 *⑤

Since 5 is the only common factor, it is also the greatest common factor (GCF).

GCF = 5.

Note:

➢ The GCF between a small and big numbers will be a small number if the big number can be divided by the small number without remaining. For example, the GCF between 3 and 12 is 3 because 12 can be divided by 3 without remaining.

➢ If the big number cannot be divided by the small number without remaining, then you need to use the three steps of finding the GCF.

Example 5) Find the GCF of 24, and 36.

24 = ②*②* 2 *③

36 = ②*②* 3 *③

GCF = 2 * 2 * 3 = 12.

The greatest common factor is an expression of the highest degree that divides each term of the polynomial. The variable part of the greatest common factor always contains the smallest power of a variable that appears in all terms of the polynomial. For example, the GCF between x^9 and x^{12} is x^9.

Example 6) Find the GCF of $6m^4$, $9m^2$ and $12m^5$.

$6m^4 = 2 \cdot 3 \cdot m \cdot m \cdot m \cdot m$

$9m^2 = 3 \cdot 3 \cdot m \cdot m$

$12m^5 = 2 \cdot 2 \cdot 3 \cdot m \cdot m \cdot m \cdot m \cdot m$

GCF $= 3*m*m = 3\ m^2$.

Summary

❖ **Factor** is a number that divides into a whole number without remaining.

❖ **Prime numbers** are numbers that only have two factors: one, and the number itself (Prime Number is a number that can only divided by 1 and itself without remaining).

❖ **Composite Number** is a number greater than 1 with more than two factors.

❖ **Prime Factorization** is a composite number expressed as a product of prime numbers.

❖ **Greatest Common Factor (GCF)** is the largest factor that two or more numbers have in common.

Exercise Set 4.3

Q1) Find the GCF of 5 and 25.

Q2) Find the GCF of 7 and 49.

Q3) Find the GCF of 16 and 20.

Q4) Write the prime factorization of 75.

Q5) Write the prime factorization of 100.

Q6) Which one of the following numbers is prime or composite:

 a) 11

 b) 13

 c) 23

 d) 27

 e) 39

Q7) Find the greatest common factor (GCF) of the following monomials:
 $x^2 and\ x^3$.

Q8) Find the greatest common factor (GCF) of the following monomials:
 $15x^2 y, 30xy^2 and\ 45x^3 y^3$.

Q9) Find the GCF between 11x and 22x.

Q10) Find the GCF between $84x^2 and\ 102\ x^5 y^3$.

To solve an equation, means finding a value of a variable (or all values of a variable) that make an equation true.

Solving Linear Equations (Review Section 2.3)

Solving linear equations is undoing operations that are being done to the variable.

To solve a linear equation, do the following steps:

1) Always isolate the variable in the liner equations.
2) Put the variable alone on one side of the equal sign.
3) Move all other terms to the other side of the equation.
4) Combine the like terms.
5) Find the value of the variable.
6) Check your answer with original equation.

Key Points

1) When solving an equation, the goal is to get the variable by itself.
2) The equation is like a balance scale. Whatever you do to one side of the equation, must be done to the other side of the equation to keep it balanced.
3) In a linear equation, if a number has been added to the variable, subtract that number from both sides of the equation
4) In a linear equation, if a number has been subtracted from the variable, add that number to both sides of the equation.
5) Anytime you move a number or variable from left to right (or right to left) of the equation, you need to use the opposite sign.
6) If a variable has been multiplied by a nonzero number (has a coefficient), divide both sides by that number (coefficient).
7) If a variable has been divided by a number, multiply both sides by that number.

Example 1) Solve x + 2 = 9.

Solution:

$x + 2 = 9$

$-2 \quad -2$

> Here number 2 has been added to the variable x so we subtract 2 from both sides of the equation per Note#2 above.

$x = 7.$

There is another shortcut way to solve $x + 2 = 9$ by moving $+2$ to the right-hand side then use the opposite sign that will be negative 2 because the sign of 2 is positive per Note#5 above.

Thus, we can write

$x + 2 = 9$ as

$x = 9 - 2$

$ = 7$

Example 2) Solve x – 2 = 7.

Solution:

$x - 2 = 7$

$+2 \quad +2$

> Here 2 has been subtracted to the variable x so we add 2 to the both sides of the equation per Note#3 above.

$x = 9.$

You can also solve $x - 2 = 7$ by moving -2 to the right and change the sign to positive

$x = 7 + 2$

$x = 9$

Example 3) Solve 3x – 2 = 19.

Solution:

$3x - 2 = 19$

$3x = 19 + 2$

$3x = 21$

$$\frac{3x}{3} = \frac{21}{3}$$

$x = 7.$

Solving Quadratic Equations

A quadratic equation is an equation that can be written in standard form as $ax^2 + bx + c = 0$ where a, b, and c are real numbers with $a \neq 0$.

In other words, we define a quadratic equation as an equation that contains a variable squared in it, and no higher powers of the variable.

Note:

Equations with the highest exponent = 2 are quadratic equations.

Example 4) Find a, b, c for the following quadratic equation:

$$5x^2 + 6x + 7 = 0$$

Solution:

Note:

A quadratic equation is an equation that contains a variable squared in it, and no higher powers of the variable.

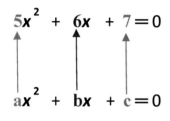

$$ax^2 + bx + c = 0$$

$a = 5$

$b = 6$

$c = 7$.

The following quadratic equations are examples of not standard and standard form:

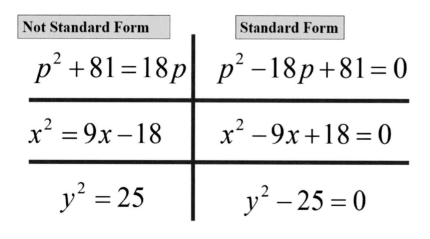

Not Standard Form	Standard Form
$p^2 + 81 = 18p$	$p^2 - 18p + 81 = 0$
$x^2 = 9x - 18$	$x^2 - 9x + 18 = 0$
$y^2 = 25$	$y^2 - 25 = 0$

Note:

A standard form of quadratic equation has a zero in one side of the equation.

The solutions of a quadratic equation in standard form $ax^2 + bx + c = 0$ with $a \neq 0$ is

$$x = \frac{-b \pm \sqrt{b^2 - 4ac}}{2a}.$$

Steps to find the solution to a quadratic equation:

1) Find the value of a, b, c.

2) Substitute a, b, c in

$$x = \frac{-b \pm \sqrt{b^2 - 4ac}}{2a}$$

3) Simplify the above equation after you have done with step 2.

4) Work it twice ±.

Example 5) Solve the following equation using the quadratic formula:

$$x^2 + 8x + 15 = 0.$$

Solution:

1) Find the value of a, b, and c.
   a = 1
   b = 8
   c = 15

2) Substitute a = 1, b = 8, c = 15 in

$$x = \frac{-b \pm \sqrt{b^2 - 4ac}}{2a}$$

$$x = \frac{-8 \pm \sqrt{8^2 - 4*1*15}}{2 \cdot 1}.$$

3) Simplify

$$x = \frac{-8 \pm \sqrt{8^2 - 4*1*15}}{2 \cdot 1}$$

$$x = \frac{-8 \pm \sqrt{64 - 60}}{2}$$

$$x = \frac{-8 \pm \sqrt{4}}{2}$$

$$x = \frac{-8 \pm 2}{2} \quad .$$

4) Work it twice ±.

$$x = \frac{-8 + 2}{2} \quad \text{or} \quad x = \frac{-8 - 2}{2}$$

$$x = \frac{-6}{2} \quad \text{or} \quad x = \frac{-10}{2}$$

$$x = -3 \quad \text{or} \quad x = -5$$

Example 6) Solve the following equation using the quadratic formula:

$$x^2 + 10x + 24 = 0 \;.$$

Solution:

a = 1

b = 10

c = 24

$$x = \frac{-b \pm \sqrt{b^2 - 4ac}}{2a}$$

$$x = \frac{-(10) \pm \sqrt{(10)^2 - 4(1)(24)}}{2(1)} = \frac{-10 \pm \sqrt{100 - 96}}{2}$$

$$x = \frac{-10 \pm \sqrt{4}}{2} = \frac{-10 \pm 2}{2}$$

$$x = \frac{-10 - 2}{2} = \frac{-12}{2} = -6$$

$$\text{or} \quad x = \frac{-10 + 2}{2} = \frac{-8}{2} = -4 \;.$$

The Zero-Product Property (Review Section 3.3)

If the product of two algebraic expressions is zero, at least one of the factors will be equal to zero.

If a and b are real numbers and if $a * b = 0$, then

$a = 0$ or $b = 0$.

> Note: If two terms multiply to zero, then either one or both terms should equal zero.

Steps of using the Zero-Product Property:

1) Set each part = 0
2) Solve each part (find the value of the variable).

Example 7) Solve the equation (x-5) * (x+2) = 0 using the Zero Product Property.

Solution:

x – 5 = 0 or x + 2 = 0

x = 5 or x = -2.

Example 8) Solve (x + 3) (x - 8) = 0.

Solution:

x + 3 = 0 or x - 8 = 0

x = -3 or x = 8.

Example 9) Solve x (x - 6) = 0.

Solution:

x = 0 or x - 6 = 0

x = 0 or x = 6.

Example 10) Solve (**4**x + 1) (**9**x - 7) = 0.

Solution:

4x + 1 = 0 or **9**x - 7 = 0

4x = -1 or **9**x = 7

$$\frac{4x}{4} = \frac{-1}{4} \qquad \text{or} \qquad \frac{9x}{9} = \frac{7}{9}$$

$$x = \frac{-1}{4} \qquad \text{or} \qquad x = \frac{7}{9}$$

The following are various factoring methods:

1) Factoring using the Greatest Common Factor (GCF)

Conditions to Factor a Polynomial Using the GCF Method

Step#1: Find the GCF between each monomial.

Step#2: Divide each monomial in the polynomial by the GCF.

Step#3: Express the polynomial as the product of the quotient and the GCF.

Example 11) Factor the following polynomial using the GCF Method:

$3x + 12$

Solution:

$3x + 12 = \mathbf{3}\,(x + 4)$

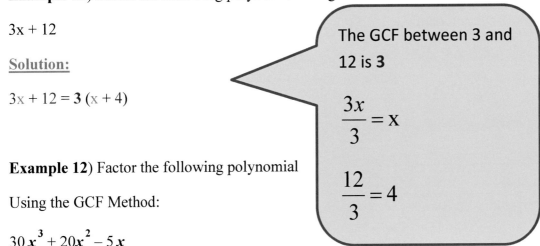

The GCF between 3 and 12 is **3**

$$\frac{3x}{3} = x$$

$$\frac{12}{3} = 4$$

Example 12) Factor the following polynomial

Using the GCF Method:

$30\,x^3 + 20x^2 - 5\,x$

Solution:

$$30\,x^3 + 20x^2 - 5\,x = 5x\,(6\,x^2 + 4x - 1)$$

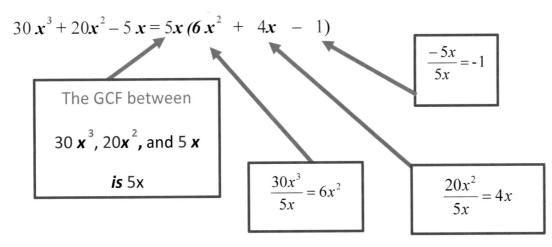

The GCF between

$30\,x^3,\ 20x^2,$ and $5\,x$

is 5x

$$\frac{30x^3}{5x} = 6x^2$$

$$\frac{20x^2}{5x} = 4x$$

$$\frac{-5x}{5x} = -1$$

2) The differences between two squares

If x and y are real numbers, variables, or algebraic expressions then,

$$x^2 - y^2 = (x - y)(x + y).$$

In other words, the difference of the squares of two terms, factors as the product of a difference and sum of those terms.

Conditions for Factoring Difference of Two Squares

Step#1: It must be a binomial (have two terms).

Stpe#2: The first term must be a perfect square.

Stpe#3: The last term must be a perfect square.

Step#4: There must be a subtraction (negative sign)

in between the two terms.

> Note: If a variable with an exponent has an even exponent then it is a perfect square.
>
> To get the square root, we simply divide the exponent by 2. For example, x^6 is a perfect square, its square root is x^3
>
> $$x^6 = (x^3)^2$$

Example 13) Factor $x^2 - 9$.

Solution:

$$x^2 - 9 = x^2 - 3^2 = (x - 3)(x + 3).$$

> Notice that:
>
> 1) $x^2 - 9$ is a binomial (stpe#1).
> 2) The first term is x^2 which is a perfect square (step#2).
> 3) The last term is 9 which is a perfect square because $9 = 3^2$ (step#3).
> 4) There is a subtraction between x^2 and 9 (step#4).

Example 14) Factor $25x^2 - 81y^2$.

Solution:

$$25x^2 - 81y^2 = (5x)^2 - (9y)^2 = (5x - 9y)(5x + 9y).$$

Here we express each term as the square of a monomial.

$$25x^2 = (5x)^2$$
$$81y^2 = (9y)^2$$

3) Factoring Perfect Square Trinomials

If x and y are real numbers, variables, or algebraic expressions then,

$$x^2 + 2x*y + y^2 = (x + y)^2$$

Same signs

and

$$x^2 - 2x*y + y^2 = (x - y)^2$$

Same signs

Conditions for Identifying a Perfect Square

Step#1: It must be a trinomial (have three terms).

Stpe#2: The first term must be a perfect square.

Stpe#3: The last term must be a perfect square.

Step#4: The middle term must be 2 or -2 times the square root of the first and last terms.

Example 15) Factor $x^2 + 8x + 16$.

Solution:

$$x^2 + 8x + 16 = (x + 4)^2.$$

Notice that $x^2 + 8x + 16$ is a perfect square because of the following steps:

1) It is a trinomial (three terms).
2) The first term is a perfect square $(x)(x) = x^2$.
3) The last term is a perfect square $(4)(4) = 16$.
4) The middle term is 2 times the square root of the first and last terms $(2)(4)(x) = 8x$.

Example 16) Factor $x^2 - 6x + 9$.

$$x^2 - 6x + 9 = (x - 3)^2.$$

Example 17) Factor $9x^2 - 30xy + 25y^2$.

$$9x^2 - 30xy + 25y^2 = (3x - 5y)^2.$$

Example 18) Factor $x^2 + 6xy + 9y^2$.

$$x^2 + 6xy + 9y^2 = (x + 3y)^2.$$

Example 19) Factor $x^2 - 8xy + 16y^2$.

$$x^2 - 8xy + 16y^2 = (x - 4y)^2$$

Example 20) Factor $x^2 - 10x + 25$.

$$x^2 - 10x + 25 = (x - 5)^2.$$

Note:

$$(x - 5)^2 = (x - 5) * (x - 5)$$

$$= x^2 - 10x + 25.$$

4) Factoring the Difference & Sum of Two Cubes

If x and y are real numbers, variables, or algebraic expressions then,

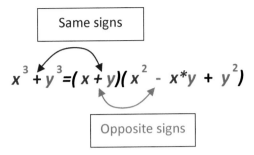

Same signs

$$x^3 + y^3 = (x + y)(x^2 - x*y + y^2)$$

Opposite signs

and

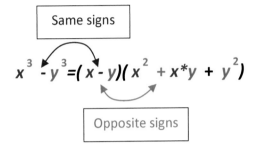

Same signs

$$x^3 - y^3 = (x - y)(x^2 + x*y + y^2)$$

Opposite signs

Note:

1) If a variable with an exponent has an exponent which is divisible by 3 without remaining, then it is a perfect cube.
2) To get the cube root, we simply divide the exponent by 3. For example, x^{12} is a perfect cube, its cube root is x^4.

$$x^{12} = (x^4)^3.$$

Conditions for Factoring Difference of Two cubes

Step#1: It must be a binomial (have two terms).

Stpe#2: The first term must be a perfect cube.

Stpe#3: The last term must be a perfect cube.

Step#4: There must be a subtraction (negative sign) in between the two terms.

Example 21) Factor $a^3 - 8$.

Solution:

$$a^3 - 8 = a^3 - 2^3 = (a - 2)(a^2 + a*2 + 2^2)$$

$$= (a - 2)(a^2 + 2a + 4).$$

Example 22) Factor $x^3 - 125$.

Solution:

$$x^3 - 125 = x^3 - 5^3 = (x - 5)(x^2 + x*5 + 5^2)$$

$$= (x - 5)(x^2 + 5x + 25).$$

Conditions for Factoring Sum of Two Cubes

Step#1: It must be a binomial (two terms).

Stpe#2: The first term must be a perfect cube.

Stpe#3: The last term must be a perfect cube.

Step#4: There must be an addition (positive sign) in between the two terms.

Example 23) Factor $x^3 + 64$.

Solution:

$$x^3 + 64 = x^3 + 4^3 = (x + 4)(x^2 - x*4 + 4^2)$$

$$= (x + 4)(x^2 - 4x + 16).$$

Example 24) Factor $27x^3 + 1$.

Solution:

$$27x^3 + 1 = (3x)^3 + 1^3 = (3x + 1)(9x^2 - 3x*1 + 1^2)$$
$$= (3x + 1)(9x^2 - 3x + 1).$$

Conditions for Factoring Trial and Error Method

Step#1: It must be a trinomial (three terms).

Stpe#2: The first and last term must be a perfect square.

Stpe#3: The coefficient of the 1st term must be one.

Step#4: After factoring out the common factor, the trinomial will factor into two binomials (two terms) as shown below (if it is factorable):

> Factoring a polynomial means expressing it as a product of other polynomials.

$$(\text{First term} \pm \text{Last term})\ (\text{First term} \pm \text{Last term})$$

Notes:

1) The product of the first terms always multiply together to give you the first term of the trinomial.

2) The product of the last terms always multiply together to give you the last term of the trinomial.

3) If the sign of the last term of the trinomial is positive then the signs will be the same as the middle term either positives or negatives. If the last term of the trinomial is negatives then the signs will be different, one positive and one negative.

4) Be sure to include the signs of the last terms when multiplying and adding.

Example 25) Factor $x^2 + 9x + 20$.

<u>Solution:</u>

We can use trial and error method since this is a trinomial and the coefficient of the first term (x^2) is one. We begin by setting up the first terms as follows:

$$x^2 + 9x + 20 = (x \quad)(x \quad).$$

Next we get the signs right. Since the last term is positive, the signs will be the same as the middle term which is positive $(x + \quad)(x + \quad)$.

Finally, we determine the last term by trial and error. The last terms multiplied together should equal **20** and when added should equal **9**. Therefore, our last terms will be **+5** and **+4**.

$$(x + 5)(x + 4)$$

$$x^2 + 9x + 20 = (x + 5)(x + 4).$$

Example 26) Factor $x^2 - 5x - 24$.

<u>Solution:</u>

We can use trial and error method since this is a trinomial and the coefficient of the first term (x^2) is one. We begin by setting up the first terms as follows:

$$x^2 - 5x - 24 = (x \quad)(x \quad).$$

Next we get the signs right. Since the last term is negative, one term will be positive and the other will be negative. $(x \quad - \quad)(x \quad + \quad)$.

Finally, we determine the last term by trial and error.

The last terms multiplied together should equal $- 24$ and when added should equal $- 5$. Therefore, our last term will be $- 8$ and $+3$.

$$(x \quad - \quad 8)(x \quad + 3 \quad)$$

$$x^2 - 5x - 24 = (x \quad - \quad 8)(x \quad + 3 \quad).$$

Example 27) Factor $x^2 - 14x + 24$.

<u>Solution:</u>

$$x^2 - 14x + 24 = (x \quad - \quad 12)(x \quad + 2 \quad).$$

> **Note:** List the factors of 24.
>
> $24 = 1 * 24$
>
> $24 = 2 * 12$
>
> $24 = 3 * 8$
>
> $24 = 4 * 6$
>
> We choose 2 and 12 because the sum of 2 and 12 is 14 which is the middle term.

Example 28) Factor $x^2 - 5x - 24$.

<u>Solution:</u>

$$x^2 - 5x - 24 = (x \quad - \quad 8)(x \quad + 3).$$

Example 29) Factor $x^2 + 7x + 12$.

Solution:

$$x^2 + 7x + 12 = (x + 4)(x + 3).$$

6) Factoring Trinomials by the Trial and Error Method when the coefficient of the 1st term is not one

Conditions for Factoring Trial and Error Method

Step#1: It must be a trinomial (three terms).

Stpe#2: The first and last term must be a perfect square.

Stpe#3: The coefficient of the 1st term is not one.

Step#4: After factoring out the common factor, the trinomial will factor into two binomials (two terms) as shown below (if it is factorable):

(First term ± Last term) (First term ± Last term)

Notes:

1) Multiply the first terms to give you the first term of the trinomial.
2) Multiply the last terms to give you the last term of the trinomial.
3) If the sign of the last term of the trinomial is positive, then the signs will be the same as the middle term either positives or negatives. If the last term of the trinomial is negatives then the signs will be different, one positive and one negative.
4) Be sure to include the signs of the last terms when multiplying and adding.

Example 30) Factor $6x^2 - 13x - 5$.

Solution:

$$6x^2 - 13x - 5 = (2x - 5)(3x + 1).$$

Example 31) Factor $12y^2 + 11y - 5$.

Solution:

$$12y^2 + 11y - 5 = (4y + 5)(3y - 1).$$

Conditions for Factoring by Grouping

Step#1: A polynomial must be four or more terms.

Step#2: Appolynomial may contain any two like terms that have anything in common, called the greatest common factor or GCF. If so, factor out the GCF.

Strategy to Factor by Grouping

1) Group it into two groups.
2) Put any two like terms in the first group in parentheses and the other two like terms in the 2nd group in parentheses.
3) Factor the first two terms using GCF method.
4) Factor the last two terms (group 2) using the GCF Method.
5) Factor out the GCF again (the GCF is the factor from step 3 & 4).
6) Check by multiplying using FOIL method.

Example 32) Factor xy + 2x + 4y + 8.

<u>Solution:</u>

Since this is a polynomial with four terms, we can factor it by grouping.

We can put the first two terms (xy and 2x) in parantheses (first group) and the last two terms (4y and 8) in another paranteheses (2nd group) as in the following:

xy + 2x + 4y + 8 = (xy + 2x) + (4y + 8)

= x (y + 2) + 4(y + 2)

= (y + 2) (x + 4).

Group it into two groups.
We factor the first two terms:
xy + 2x = x (y + 2).
and we factor the last two terms:
4y + 8 = 4(y + 2).

Here we factor it out again
y + 2 is the GCF between
x (y + 2) and 4(y + 2).

Example 33) Factor $3xy + 2x + 21y + 14$.

Solution:

$3xy + 2x + 21y + 14 = (3xy + 2x) + (21y + 14)$ ⟵ Group it into two groups.

$= x(3y + 2) + 7(3y + 2)$ ⟵ Factor out GCF from each group

$= (3y + 2)(x + 7)$. ⟵ Factor out the GCF again

Example 34) Factor $x^3 - 3x^2 + 4x + 12$.

Solution:

$x^3 - 3x^2 + 4x - 12 = (x^3 - 3x^2) + (4x - 12)$

$= x^2(x-3) + 4(x - 3)$

$= (x-3)(x^2 + 4)$.

Solving Quadratic Equations by Factoring

Strategy for Solving a Quadratic Equation by Factoring

Step #1) Write the equation in standard form $ax^2 + bx + c$.

Step #2) Factor the equation completely.

Step #3) Use the zero-factor property.

Set each factor with a variable $= 0$.

Step #4) Solve each equation produced in step #3.

Step #5) Check each solution in the original equation.

Example 35) Solve $x^2 - 7x = 8$.

Solution:

Step #1) Write the equation in standard form $ax^2 + bx + c$.

$x^2 - 7x = 8$

$x^2 - 7x - 8 = 0$.

Step #2) Factor completely

$x^2 - 7x - 8 = 0$.

$(x - 8)(x + 1) = 0$

$(x - 8) = 0$ or $(x + 1) = 0$.

Step #3) Use the zero-factor property. Set each factor with a variable = 0.

Step #4) Solve each equation produced in step #3.

$x - 8 = 0$ or $x + 1 = 0$

$x = 8$ or $x = -1$.

Step #5) Check each solution in the original equation.

$x = 8$	Or	$x = -1$
$x^2 - 7x = 8$		$x^2 - 7x = 8$
$8^2 - 7(8) = 8$		$(-1)^2 - 7(-1) = 8$
$64 - 56 = 8$		$1 + 7 = 8$
$8 = 8$.		$8 = 8$.

Example 36) Solve $x^2 = 9x - 18$.

Solution:

$x^2 = 9x - 18$

$x^2 - 9x + 18 = 0$

$(x - 6)(x - 3) = 0$

$x - 6 = 0$ or $x - 3 = 0$

$x = 6$ or $x = 3$.

Notes:

1) If it is a quadratic equation, you should always change it to standard form.
2) If an equation has a degree of 2 or higher, we cannot solve it until it has been factored.

Example 37) Solve $x^2 = -2x + 3$.

Solution:

$x^2 = -2x + 3$

$x^2 + 2x - 3 = 0$

$(x + 3)(x - 1) = 0$

$x + 3 = 0$ or $x - 1 = 0$

$x = -3$ or $x = 1$.

Example 38) Solve $x^2 = 5x$.

Solution:

$x^2 = 5x$

$x^2 - 5x = 0$

$x(x - 5) = 0$ here x is the GCF

$x = 0$ or $x - 5 = 0$

$x = 0$ or $x = 5$.

Example 39) Find two consecutive integers whose product is 210.

Let x = 1st integer.

Let x + 1 = 2nd integer.

x (x + 1) = 210

$x^2 + x = 210$

$x^2 + x - 210 = 0$

(x + 15) (x -14) = 0

x +15 = 0   or x -14 = 0

x = -15   or x = 14

The consecutive integers are 14, 15 or -14, -15.

Summary

To factor an expression means to write an equivalent expression that is a product (multiplication). Factoring is a method to find the basic numbers and variables that made up a product. A factor that cannot be factored further is said to be a prime factor (in other words, some numbers are Prime, meaning they are only divisible without remaining by themselves and 1 and not factorable.
A polynomial is factored completely if it is written as a product of prime polynomials.

To factor any polynomials,

1) Always look for a GCF before using any other factoring methods.
2) Count the number of terms in a polynomial. If it is:
   a. 4 terms, look to factor it by grouping.
   b. 3 terms, look to factor it as perfect square trinomial. If not, look to factor it as trial and error method.
   c. 2 terms, try to factor it by difference of two squares if the first and second terms are perfect squares. If the first and second terms are perfect cubes, look to factor it out by difference or sum of two cubes.

The following table contains a list of perfect squares and cubes:

List of perfect squares	List of perfect cubes
$1^2 = 1$	$1^3 = 1$
$2^2 = 4$	$2^3 = 8$
$3^2 = 9$	$3^3 = 27$
$4^2 = 16$	$4^3 = 64$
$5^2 = 25$	$5^3 = 125$
$6^2 = 36$	$6^3 = 216$
$7^2 = 49$	$7^3 = 343$
$8^2 = 64$	$8^3 = 512$
$9^2 = 81$	$9^3 = 729$
$10^2 = 100$	$10^3 = 1000$

Note: Anytime you see the word "solve", it means find the value of a variable in the equation.

To solve (find the value of x) the equation $x - 2 = 3$,

$x - 2 = 3$

$x = 3 + 2 = 5$.

Exercise Set 4.4

Q1) Solve $2x - 3 = 9$.

Q2) solve $x(x-15) = 0$.

Q3) Solve the following quadratic equation:

$6x^2 + 5x = 4$.

Q4) Factor the following by using the grouping method:

$3x^2 - 9x + 5x - 15$

Q5) $(x - 3)(x + 5) = 0$.

Q6) Factor the following:

    a) $5x + 45$

    b) $x^2 - 64$

    c) $16x^2 - 49y^2$

    d) $x^2 + 10x + 25$

    e) $x^2 - 14x + 49$

    f) $x^2 - 5x + 6$

    h) $3x^2 + 11x - 20$

    i) $x^2 + 10x + 21$

    j) $y^2 + 8y + 15$

    k) $y^2 - 11y + 24$

Q7) The square of a number minus twice the number is 63. Find the number.

Q8) The difference of 10 and a number is negative eight. Find the number.

Q9) The length of a rectangular garden is 10 feet more than its width. The area of the garden is 600 square feet. What are the length and the width of the garden?

Q10) Nancy bought a pair of skirts at the mall and 3 shorts for $45 at the clothing store. All together she spent $85 for the clothes. How much was each skirt?

Q11) What are the solutions to the following quadratic equation

    $x^2 - 2x - 48 = 0$?

  A. 6 and 8

  B. -6 and -8

  C. -6 and 8

  D. 6 and -8

  E. 3 and 16

Q12) What is the sum of the solutions to the quadratic equation $x^2 + 12x = 0$?

Chapter 5

Radicals, Distance, and Midpoint Formulas

Section 5.1 Introduction to Radicals, Adding and Subtracting Radical Expressions

Introduction to Radicals

A radical is an expression that has a square root, cube root, etc. The three components of a radical are:

1) **Index of a radical (a radical degree):** The number of times the radicand is multiplied by itself for example, 2 means square root, 3 means cube root. After that they are called the 4th root, 5th root and so on. If this is missing, it is assumed to be 2 (the square root).

2) **Radical symbol:** $\sqrt[n]{}$ is the radical symbol that means "nth root of".

3) **Radicand:** A term or expression under the radical symbol. Below is an example of a radical with its three components:

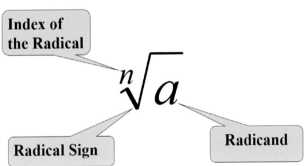

Index of the Radical

$\sqrt[n]{a}$

Radical Sign

Radicand

Perfect Squares

The Perfect Squares (Square Numbers) are the squares of the whole numbers. The following are some lists of perfect squares:

$1^2 = 1$	$5^2 = 25$	$9^2 = 81$
$2^2 = 4$	$6^2 = 36$	$10^2 = 100$
$3^2 = 9$	$7^2 = 49$	$11^2 = 121$
$4^2 = 16$	$8^2 = 64$	$12^2 = 144$

Square Root

A square root of a number or variable is a term that can be multiplied by itself to give the original number or variable. If a variable with an exponent has an even exponent, then it is a perfect square. To get the square root, we simply divide the exponent by 2 as in the following example:

$$\sqrt{x^{10}} = x^5$$

$10 \div 2 = 5$ or we can write
$\sqrt{x^{10}} = x^5$ bcause $x^5 * x^5 = x^{10}$

Below is a list of square roots of some perfect square numbers:

$\sqrt{0} = 0$	$\sqrt{36} = 6$	$\sqrt{144} = 12$
$\sqrt{1} = 1$	$\sqrt{49} = 7$	$\sqrt{169} = 13$
$\sqrt{4} = 2$	$\sqrt{64} = 8$	$\sqrt{196} = 14$
$\sqrt{9} = 3$	$\sqrt{81} = 9$	$\sqrt{225} = 15$
$\sqrt{16} = 4$	$\sqrt{100} = 10$	$\sqrt{400} = 20$
$\sqrt{25} = 5$	$\sqrt{121} = 11$	$\sqrt{625} = 25$

Perfect Cubes

Perfect cubes are the cubes of the whole numbers.

The following is a list of perfect Cubes

$1^3 = 1$
$2^3 = 8$
$3^3 = 27$
$4^3 = 64$
$5^3 = 125$
$6^3 = 216$
$7^3 = 343$
$8^3 = 512$
$9^3 = 729$
$10^3 = 1000$

Life is a short trip. Enjoy it as much as you can before the trip is over!

Dilshad Akrayee

If a variable with an exponent has an exponent which is divisible by 3 then it is a perfect cube. To get the cube root, we simply divide the exponent by 3 as in the following example:

$$\sqrt[3]{x^{15}} = x^5$$

$15 \div 3 = 5$ or we can write:
$\sqrt[3]{x^{15}} = x^5$ because $x^5 * x^5 * x^5 = x^{15}$

Below is a list of cube roots of some perfect cubes numbers:

$\sqrt[3]{0} = 0$	$\sqrt[3]{64} = 4$	$\sqrt[3]{512} = 8$
$\sqrt[3]{1} = 1$	$\sqrt[3]{125} = 5$	$\sqrt[3]{729} = 9$
$\sqrt[3]{8} = 2$	$\sqrt[3]{216} = 6$	$\sqrt[3]{1000} = 10$
$\sqrt[3]{27} = 3$	$\sqrt[3]{343} = 7$	$\sqrt[3]{1331} = 11$

Adding and Subtracting Radical Expressions

To combine like radicals (with the same index and radicand), combine the coefficients of these radicals.

Like Radicals and Unlike Radicals

Radicals can be of two types as in the following:

1) Like Radicals
2) Unlike Radicals

Radicals with the same index and radicand are called like radicals. Below is a list of like radicals:

Like Radicals (Square Roots)	Like Radicals (Cube Roots)
$\sqrt{5}$ and $3\sqrt{5}$	$24\sqrt[3]{7}$ and $19\sqrt[3]{7}$
$2\sqrt{x}$ and $5\sqrt{x}$	$-4\sqrt[3]{xy}$ and $15\sqrt[3]{xy}$
$3\sqrt{2}$, $4\sqrt{2}$, and $-7\sqrt{2}$	$8\sqrt[3]{7}$, $-11\sqrt[3]{7}$, and $10\sqrt[3]{7}$
$-4\sqrt{x^2y^5}$, $5\sqrt{x^2y^5}$, $7\sqrt{x^2y^5}$, and $\sqrt{x^2y^5}$	$5\sqrt[3]{xy^2z^5}$, $-11\sqrt[3]{xy^2z^5}$, $9\sqrt[3]{xy^2z^5}$ and $10\sqrt[3]{xy^2z^5}$

Notes:

1) Only like radicals can be combined.
2) Radicals are just like fractions that need common denominators, indexes, and radicands when adding/subtracting.
3) You may need to convert radicals to a different form (entire or mixed) before identifying like radicals.

Example 1) Add the following:

$$3\sqrt{7} + 6\sqrt{7}$$

Solution:

$$3\sqrt{7} + 6\sqrt{7} = 9\sqrt{7}$$

> Here $3\sqrt{7}$ *and* $6\sqrt{7}$ have the same index which is 2 and have the same radicand which is 7 so we combine their coefficients $3 + 6 = 9$.

Example 2) Subtract

$$10\sqrt[3]{x} - 6\sqrt[3]{x}$$

Solution:

$$10\sqrt[3]{x} - 6\sqrt[3]{x} = 4\sqrt[3]{x}$$

> Here $10\sqrt[3]{x}$ *and* $6\sqrt[3]{x}$ have the same index which is 3 and have the same radicand which is x so we subtract their coefficients $10 - 6 = 4$.

Example 3) Add

$$2\sqrt{5} + 4\sqrt{5}$$

Solution:

$$2\sqrt{5} + 4\sqrt{5} = 6\sqrt{5}$$

> Here we combine the like radicals by adding the numbers multiplied by the radical.
>
> $2 + 4 = 6$.

Example 4) Subtract

$$6\sqrt{x} - 2\sqrt{x}$$

Solution:

$$6\sqrt{x} - 2\sqrt{x} = 4\sqrt{x}$$

> Here we combine the like radicals by subtracting the numbers multiplied by the radical.
>
> $6 - 2 = 4$.

Unlike Radicals

Radicals with the different index are called unlike radicals. Radical Expressions will be unlike radicals if:

1) We have two radical expressions such as 'A' and 'B', when 'A' and 'B' are radicands and they have different index over the radical sign.
2) Two radical expressions have radicands such as 'A' and 'B' and $A \neq B$.

Below is a list of unlike radicals:

Unlike Radicals
$\sqrt[3]{7}$ and $4\sqrt{5}$
$2\sqrt{x}$ and $5\sqrt{xy}$
$3\sqrt{2}$, $4\sqrt[5]{2}$,
$-4\sqrt{x^2y^2}$ and $5\sqrt{x^3y^5}$

Exercise Set 5.1

Q1) Add $\sqrt{5} + 3\sqrt{5}$.

Q2) Find the perimeter of the following triangle. Give the answer as a radical expression in simplest form.

Q3) Add $2\sqrt{x} + 5\sqrt{x}$.

Q4) Add $4\sqrt[3]{7} + 5\sqrt[3]{7}$.

Q5) $\sqrt[3]{216} = ?$

Q6) Subtract $28\sqrt[7]{xyz} - 11\sqrt[7]{xyz}$.

Q7) Subtract $8\sqrt[3]{a} - 11\sqrt[3]{a}$.

Q8) Add or subtract $2\sqrt[5]{x} + 7\sqrt[5]{x} - 11\sqrt[5]{x}$.

Q9) Subtract $5\sqrt[7]{x^2y^5} - 3\sqrt[7]{x^2y^5}$.

Q10) $\sqrt[5]{x^{15}} = ?$

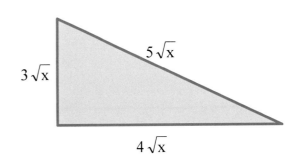

Radicals have the following properties:

    1) Product Property of Radicals

If $\sqrt[n]{x}$ and $\sqrt[n]{y}$ are any real numbers and "n" is any integer where n > 1 then

$$\sqrt[n]{x} * \sqrt[n]{y} = \sqrt[n]{x * y}.$$

In other words, to multiply radicals do the following steps:

    Step#1) Multiply the coefficients (the numbers or variables on the outside).

    Steps#2) Multiply the radicands (the numbers or variables on the inside).

    Step# 3) Simplify the remaining radicals.

Example #1) Multiply

$3\sqrt{2} * 4\sqrt{8}.$

<u>Solution:</u>

$3\sqrt{2} * 4\sqrt{8} = 12 * \sqrt{16}$

$\qquad\qquad = 12 * 4$

$\qquad\qquad = 48.$

> Here we do the following steps:
>
>   **1)** Multiply the coefficients
>
> $\qquad\qquad 3*4 = 12.$
>
>   **2)** Multiply the radicands
>
> $\qquad\qquad 2 * 8 = 16.$
>
>   **3)** Simplify $\sqrt{16} = 4$.
>
>   **4)** Multiply $12 * 4 = 48.$

Example #2) Multiply

$2\sqrt{5} * 3\sqrt{5}.$

<u>Solution:</u>

$2\sqrt{5} * 3\sqrt{5} = 6 * \sqrt{25} = 6 * 5 = 30.$

Example #3) Multiply

$\sqrt[3]{9} * \sqrt[3]{3}.$

Solution:

$\sqrt[3]{9} * \sqrt[3]{3} = \sqrt[3]{27} = 3.$

Example #4) Multiply

$-7\sqrt[5]{x} * 3\sqrt[5]{y}.$

Solution:

$-7\sqrt[5]{x} * 3\sqrt[5]{y} = -21\sqrt[5]{xy}.$

---

2) **Quotient Property of Radicals**

If $\sqrt[n]{x}$ and $\sqrt[n]{y}$ are any real numbers and "n" is any integer where n > 1 then,

$$\frac{\sqrt[n]{x}}{\sqrt[n]{y}} = \sqrt[n]{\frac{x}{y}}.$$

In other words, the nth root of a quotient is the quotient of nth roots, and the quotient of two nth roots is the nth root of the quotient. To divide radicals, do the following steps:

Step#1) Divide the coefficients (the numbers or variables on the outside).

Steps#2) Divide the radicands (the numbers or variables on the inside).

Step# 3) Simplify the remaining radicals.

Example 5) Divide

$\dfrac{18\sqrt[3]{16}}{6\sqrt[3]{2}}.$

Solution:

$\dfrac{18\sqrt[3]{16}}{6\sqrt[3]{2}} = 3 * \sqrt[3]{8} = 3 * 2 = 6.$

Here we simplify as in the following steps:
1) Divide the coefficients
$18 \div 6 = 3$
2) Divide the radicands
$16 \div 2 = 8$
3) $\sqrt[3]{8} = 2$ becuase $2 * 2 * 2 = 8$
4) $3 * 2 = 6.$

Example 6) Divide

$$\frac{\sqrt[5]{x^4 y^7}}{\sqrt[5]{x^3 y^3}}.$$

Solution:

$$\frac{\sqrt[5]{x^4 y^7}}{\sqrt[5]{x^3 y^3}} = \sqrt[5]{\frac{x^4 y^7}{x^3 y^3}} = \sqrt[5]{xy^4}.$$

Here $\dfrac{x^4 y^7}{x^3 y^3} = x^{4-3} * y^{7-3} = x^1 y^4 = xy^4$

3) For any real numbers "a" and "x", for which the **nth** root of x is defined and any integer n>1 then,

$$\sqrt[n]{x^a} = x^{\frac{a}{n}}.$$

Example 7) $\sqrt[3]{x^6} = ?$

Solution:

$$\sqrt[3]{x^6} = x^{\frac{6}{3}} = x^2.$$

Note:

If x is any real number then,

$$\sqrt{x} * \sqrt{x} = x.$$

Such as in the following examples:

$$\sqrt{3} * \sqrt{3} = 3$$

$$\sqrt{5} * \sqrt{5} = 5$$

$$\sqrt{y} * \sqrt{y} = y.$$

Online Resources:

http://www2.highlands.edu/site/faculty-dilshad-akrayee

Definition of the Square Root of a Number

A square root of a nonnegative real number "a" is a nonnegative real number "x" such that $x^2 = a$, denoted by $x = \pm\sqrt{a}$ is the principal square root of a.

For example, if $x^2 = 9$ then $x = \pm\sqrt{9} = \pm 3$

Definition of the Cub Root of a Number

A cube root of a number "y" is a number "b" such that $\sqrt[3]{y} = b$. In other words,

if $\sqrt[3]{y} = b$ means that $b^3 = y$ and vice a verse.

For example,

if $\sqrt[3]{4} = b$ then $b^3 = 4$ and

if $b^3 = 27$ then $b = \sqrt[3]{27} = 3$ because $3*3*3 = 27$

Rationalizing the Denominator

Rationalizing the denominator is the process by which a fraction containing radicals in the denominator is rewritten to have only rational numbers in the denominator.

How to Rationalize the Denominator?

1) Multiply Both Top and Bottom by a Root.
2) Multiply Both Top and Bottom by the Conjugate. The conjugate is where we change the sign in the middle of two terms as below:

Expression Examples	It's Conjugate
$\sqrt{x} - 2$	$\sqrt{x} + 2$
$3 - \sqrt{5}$	$3 + \sqrt{5}$
$\sqrt{x} + \sqrt{y}$	$\sqrt{x} - \sqrt{y}$

Example 8) Rationalize the denominator of the following:

$$\frac{3}{\sqrt{2}}.$$

<u>Solution:</u>

$$\frac{3}{\sqrt{2}} = \frac{3}{\sqrt{2}} * \frac{\sqrt{2}}{\sqrt{2}} = \frac{3\sqrt{2}}{\sqrt{4}} = \frac{3\sqrt{2}}{2}.$$

Here multiply top and bottom of the fraction $\dfrac{3}{\sqrt{2}}$ by the square root of 2, because $\sqrt{2} * \sqrt{2} = \sqrt{4} = 2$

Example 9) Rationalize the denominator of the following:

$$\frac{5}{\sqrt{x}}.$$

<u>Solution:</u>

$$\frac{5}{\sqrt{x}} = \frac{5}{\sqrt{x}} * \frac{\sqrt{x}}{\sqrt{x}} = \frac{5\sqrt{x}}{\sqrt{x^2}} = \frac{5\sqrt{x}}{x}.$$

Here multiply top and bottom of the fraction $\dfrac{5}{\sqrt{x}}$ by the square root of x, because $\sqrt{x} * \sqrt{x} = \sqrt{x^2} = x$

Example 10) Rationalize the denominator of the following:

$$\frac{7}{4 - \sqrt{x}}$$

<u>Solution:</u>

$$\frac{7}{4 - \sqrt{x}} = \frac{7}{4 - \sqrt{x}} * \frac{4 + \sqrt{x}}{4 + \sqrt{x}} = \frac{7(4 + \sqrt{x})}{16 - \sqrt{x^2}} = \frac{28 + 7\sqrt{x}}{16 - x}$$

Here multiply top and bottom of the fraction $\dfrac{7}{4 - \sqrt{x}}$ by the conjugate $4 + \sqrt{x}$.

Exercise Set 5.2

Q1) Multiply $2\sqrt{5} * 4\sqrt{3}$.

Q2) Multiply $-2\sqrt{x} * 9\sqrt{yz}$.

Q3) Multiply $\sqrt[5]{A^2 B^2} * \sqrt[5]{AB}$.

Q4) Divide

$$\frac{\sqrt[7]{x^4 y^5}}{\sqrt[7]{x\, y^2}}.$$

Q5) $\sqrt{x^{10}} = ?$

Q6) Use rational exponents to simplify: $\sqrt[6]{x^3}$.

Q7) Rationalize the denominator of the following:

$$\frac{1}{5-\sqrt{3}}.$$

Q8) Rationalize the numerator of the following:

$$\frac{5-\sqrt{2}}{2}.$$

Q9) Solve $4x^2 = 81$.

Q10) If $x^3 = -125$ then x = ?

Q11) If $x^2 = \dfrac{16}{49}$ then x = ?

Q12) Rationalize the denominator of the following rational expression:

$$\frac{5-\sqrt{x}}{7+\sqrt{x}}.$$

When simplifying a radical expression, find the factors that are to the "n^{th}" powers of the radicand and then use the Product Property of Radicals (Page 138).

Simplify Radicals Involving Variables.

The square root of a squared number is always positive. The absolute value is used to express this. The product and quotient rules apply when variables appear under the radical sign, given that the variables represent only nonnegative real numbers.

For any real number a, $\sqrt{a^2} = |a|$.

Where, $a \geq 0$, $|a| = a$.

Simplifying Radical Expressions by Factoring

A radical expression whose index is "n" is simplified when its radicand has no factors that are perfect nth powers. To simplify, use the following procedures:

1) Write the radicand as the product of two factors, one of which is the greatest perfect nth power.

2) Use the product rule to take the nth root of each factor.

3) Find the nth root of the perfect nth power.

Example 1) Simplify the following radical by factoring:

$$\sqrt{28}.$$

Here we use procedures #1 and 2 above by doing the following:

1) We write the radicand 28 as the product of two factors 4 and 7. $28 = 4 * 7$ and we use the product rule.

2) $\sqrt{28} = \sqrt{4} * \sqrt{7}$.

Solution:

$$\sqrt{28} = \sqrt{4} * \sqrt{7}$$
$$= 2 * \sqrt{7}$$
$$= 2\sqrt{7}.$$

$\sqrt{4} = 2$ becuase $2*2 = 4$

Example 2) Simplify the following radical expression by factoring:

$$\sqrt{20x^2 y}.$$

<u>Solution:</u>

$$\sqrt{20x^2 y} = \sqrt{20} * \sqrt{x^2} * \sqrt{y}$$

$$= \sqrt{4} * \sqrt{5} * x * \sqrt{y}$$

$$= 2x\sqrt{5y}.$$

$\sqrt{x^2} = x$ becuase $x * x = x^2$

Example 3) Simplify the following radical expression by factoring:

$$\sqrt[3]{\frac{16x^6}{125}}.$$

<u>Solution:</u>

$$\sqrt[3]{\frac{16x^6}{125}} = \frac{\sqrt[3]{8} * \sqrt[3]{2} * \sqrt[3]{x^6}}{\sqrt[3]{125}} = \frac{2 * \sqrt[3]{2} * x^2}{5} = \frac{2\sqrt[3]{2} \, x^2}{5}.$$

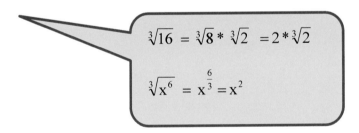

$$\sqrt[3]{16} = \sqrt[3]{8} * \sqrt[3]{2} = 2 * \sqrt[3]{2}$$

$$\sqrt[3]{x^6} = x^{\frac{6}{3}} = x^2$$

Steps for Solving Radical Equations

The following are general steps for solving radical equations:

1) Isolate one radical on one side of equal sign.

2) Raise each side of the equation to a power equal to the index of the isolated radical, and simplify. (With square roots, the index is 2, so square both sides).

3) If equation still contains a radical, repeat steps 1 and 2. If not, solve equation.

4) Check proposed solutions in the original equation.

Example 4) Solve the following radical equation:

$$\sqrt{x-1} - 2 = 1.$$

<u>Solution:</u>

$$\sqrt{x-1} - 2 = 1$$

$$\sqrt{x-1} = 1 + 2$$

> Here we isolate the radical sign by moving -2 from left side to the right side of the equation (step 1).

$$\sqrt{x-1} = 3$$

> Here we raise each side of the equation to a power 2 (step 2).

$$(\sqrt{x-1})^2 = 3^2$$

$$x - 1 = 9$$

$$x = 9 + 1$$

$$x = 10.$$

Reminder

➢ Solving radical equations requires applying the rules of exponents.
➢ A mutual method for solving radical equations is to raise both sides of an equation to whatever power will remove the radical sign from the equation.

The solution is x = 10

Substitute x = 10 in the following original equation:

$$\sqrt{x-1} - 2 = 1$$

$$\sqrt{10-1} - 2 = 1$$

$$\sqrt{9} - 2 = 1$$

$$3 - 2 = 1$$

$$1 = 1 \text{ true.}$$

Simplifying Radical Expressions (General Notes)

1) If a radical represents a rational number (fraction), use that rational number in place of the radical such as the following examples:

$$\sqrt{36} = 6, \quad \sqrt{\frac{49}{81}} = \frac{\sqrt{49}}{\sqrt{81}} = \frac{7}{9}$$

2) If a radical expression contains products of radicals, use the product property for radicals such as the following examples:

$$\sqrt{7x} = \sqrt{7} * \sqrt{x} \quad \text{and} \quad \sqrt[3]{x^2 y} = \sqrt[3]{x^2} * \sqrt[3]{y}$$

3) If a radicand of a square root radical has a factor that is a perfect square, express the radical as the product of the positive square root of the perfect square and the remaining radical factor. A similar statement applies to nth roots such as the following examples:

$$\sqrt{20} = \sqrt{4*5} = \sqrt{4} * \sqrt{5} = 2\sqrt{5}$$

$$\sqrt[3]{16} = \sqrt[3]{8*2} = \sqrt[3]{8} * \sqrt[3]{2} = 2\sqrt[3]{2}$$

4) Rationalize any dominator containing a radical such as the following examples:

$$\frac{5}{\sqrt{3}} = \frac{5 * \sqrt{3}}{\sqrt{3} * \sqrt{3}} = \frac{5\sqrt{3}}{3}$$

$$\sqrt{\frac{3}{2}} = \frac{\sqrt{3} * \sqrt{2}}{\sqrt{2} * \sqrt{2}} = \frac{\sqrt{6}}{2}$$

$$\sqrt[3]{\frac{1}{4}} = \frac{\sqrt[3]{1}}{\sqrt[3]{4}} = \frac{\sqrt[3]{1}}{\sqrt[3]{4}} * \frac{\sqrt[3]{2}}{\sqrt[3]{2}} = \frac{\sqrt[3]{2}}{\sqrt[3]{8}} = \frac{\sqrt[3]{2}}{2}$$

5) When there is more than one term in the denominator, the process is a little tricky. You will need to multiply the numerator and denominator by the denominator's conjugate. The conjugate is the same expression as the denominator, but with the opposite sign in the middle, separate the terms such the following examples:

$$\frac{1}{5 - \sqrt{x}} = \frac{1}{5 - \sqrt{x}} * \frac{5 + \sqrt{x}}{5 + \sqrt{x}} = \frac{1(5 + \sqrt{x})}{\sqrt{25} - \sqrt{x^2}} = \frac{5 + \sqrt{x}}{5 - x},$$

$$\frac{7}{5 - \sqrt{3}} = \frac{7}{5 - \sqrt{3}} * \frac{5 + \sqrt{3}}{5 + \sqrt{3}} = \frac{7(5 + \sqrt{3})}{\sqrt{25} - \sqrt{9}} = \frac{35 + 7\sqrt{x}}{5 - 3} = \frac{35 + 7\sqrt{x}}{2}$$

Pythagorean Theorem

In any right triangle (has a **90°** angle), the square of the longer side (called the hypotenuse) is equal to the sum of the squares of other two sides (called legs).

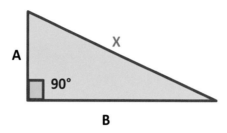

$$X = \sqrt{A^2 + B^2}.$$

Example 5) Find the distance x of the following right triangle:

Solution:

$$x = \sqrt{3^2 + 4^2}$$

$$x = \sqrt{9 + 16}$$

$$x = \sqrt{25} = 5.$$

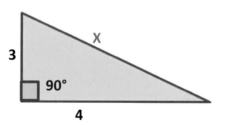

Example 6) What is the distance from point A to point B?

Solution:

$$X = \sqrt{A^2 + B^2}$$

$$= \sqrt{3^2 + 5^2}$$

$$= \sqrt{9 + 25}$$

$$= \sqrt{34}.$$

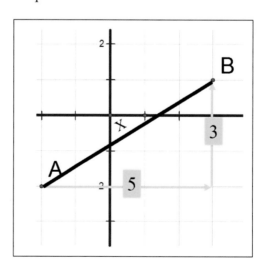

Q1) Simplify the following radical by factoring:

$\sqrt{50}$.

Q2) Simplify each of the following radicals:

a) $\sqrt{\dfrac{36}{49}}$ b) $\sqrt[3]{-8}$ c) $\sqrt[3]{\dfrac{27}{64}}$ d) $\sqrt[5]{0}$ e) $\sqrt[7]{-1}$

Q3) $\sqrt{48x^4 y^3} = ?$

Q4) Simplify

$\dfrac{\sqrt{3}}{2} + \dfrac{1}{\sqrt{3}}$.

Q5) Find the distance "d" of the following right triangle:

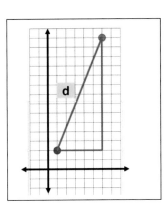

Q6) Use rational exponents to simplify: $\sqrt[6]{x^3}$.

Q7) Simplify the following radical expression using the quotient rule: $\sqrt{\dfrac{50x^3}{81y^8}}$.

Q8) Simplify by rationalizing the denominator of the following radical expression. Assume that x ≥ 0.

$\dfrac{5}{7-\sqrt{x}}$.

Q9) Solve the following radical equation:

$\sqrt{x} = 9$.

Q10) Solve the following radical equation:

$\sqrt[3]{x-2} + 3 = 6$.

Section 5.4 Distance and Midpoint Formulas

<u>Definitions</u>

Distance Formula: A formula used to find the distance between two points on a coordinate plane.

Midpoint: The points halfway between the endpoints of a segment. In other words, midpoint is a point which divides a line segment into two lines of equal length.

Segment Bisector: A segment, line, or plane that intersects a segment at its midpoint.

The Distance "d" between the points (x_1, y_1) and (x_2, y_2) is:

$$d = \sqrt{(x_2 - x_1)^2 + (y_2 - y_1)^2} \, .$$

Steps to Find the Distance Between Two Points (x_1, y_1) and (x_2, y_2)

1) First substitute numbers for variables and solve the parentheses.
2) Simplify by subtracting the numbers $(x_2 - x_1)$ and $(y_2 - y_1)$ then Solve the squared numbers.
3) Add the two numbers.
4) Find the square root of the remaining number.

Example 1) Find the distance between (2, 1) and (3, 5).

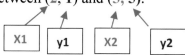

| X1 | y1 | X2 | y2 |

<u>Solution</u>

$$d = \sqrt{(x_2 - x_1)^2 + (y_2 - y_1)^2}$$

1) Substitute numbers for variables and solve the parentheses.

X1 = 2, y1 = 1, X2 = 3, and y2 = 5

$$d = \sqrt{(3-2)^2 + (5-1)^2}$$

2) Simplify by subtracting the numbers.

3 - 2 = 1 and 5 - 1 = 4

$$d = \sqrt{(1)^2 + (4)^2}$$

3) Solve the square numbers.

$(1)^2 = 1$ and $(4)^2 = 16$

$$d = \sqrt{1 + 16}$$

4) Add the two numbers

1 + 16 = 17

$$d = \sqrt{17} \approx 4.12$$

5) Find the square root of the remaining number

Example 2) Find the distance between A (5, 7) and B (2, 3).

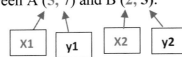

X1   y1   X2   y2

<u>Solution:</u>

$d = \sqrt{(x_2 - x_1)^2 + (y_2 - y_1)^2}$ ← 1) Substitute numbers for variables and solve the parentheses.

X1 = 5, y1 = 7, X2 = 2, and y2 = 3

$d = \sqrt{(2-5)^2 + (3-7)^2}$ ← 2) Simplify by subtracting the numbers.
2 - 5 = -3 and 3 - 7 = -4

$d = \sqrt{(-3)^2 + (-4)^2}$ ← 3) Solve the square numbers.
$(-3)^2 = 9$ and $(-4)^2 = 16$

$d = \sqrt{9 + 16}$ ← 6) Add the two numbers
9 + 16 = 25

$d = \sqrt{25} = 5.$ ← 7) Find the square root of the remaining number

Midpoint Formula

If the coordinates of the end points of a segment are (x_1, y_1) and (x_2, y_2), then the coordinates of the midpoint of this segment is given by the formula:

$$\text{Midpoint} = \left(\frac{x_1 + x_2}{2}, \frac{y_1 + y_2}{2} \right).$$

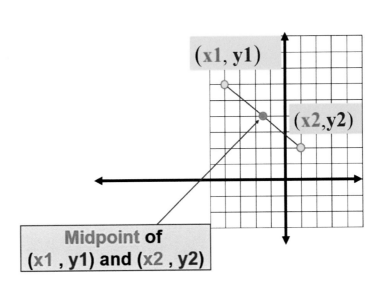

(x1, y1)

(x2,y2)

Midpoint of
(x1 , y1) and (x2 , y2)

Example 3) Find the midpoint given by the points A (-4, 6) and B (1, 2).

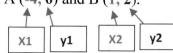

Solution:

$$\text{Midpoint} = \left(\frac{x_1 + x_2}{2}, \frac{y_1 + y_2}{2} \right)$$

$$= \left(\frac{-4+1}{2}, \frac{6+2}{2} \right) = \left(\frac{-3}{2}, 4 \right).$$

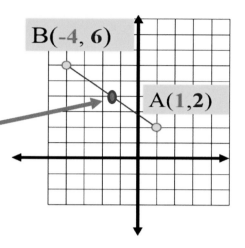

B(-4, 6)

A(1,2)

Geometry Review

What is the difference between the symbols $\overline{AB}$ and AB?

$\overline{AB}$ is the length of Segment AB.

AB is read as "line segment AB". The bar over the two letters indicates it is a line segment (Segment AB).

Example 4) Find the midpoint given by the points A (-6, 3) and B (-2, 0).

Solution:

$$\text{Midpoint} = \left(\frac{x_1 + x_2}{2}, \frac{y_1 + y_2}{2} \right)$$

$$\text{Midpoint} = \left(\frac{-6 + -2}{2}, \frac{3+0}{2} \right) = \left(-4, \frac{3}{2} \right).$$

Q1) Find the midpoint given by the points A (6, -2) and B (2, -9).

Q2) Find the midpoint given by the points A (2, 7) and B (11, 9).

Q3) Find the distance between A (4,8) and B (1,12).

Q4) Find the distance between A (-2,5) and B (3, -1).

Q5) $\overline{AB}$ has endpoints (8, 9) and (–6, –3). Find the coordinates of its midpoint M.

Q6) What is the distance from point A to point B by using the following:

   A) The Pythagorean Theorem.
   B) The Distance Formula.

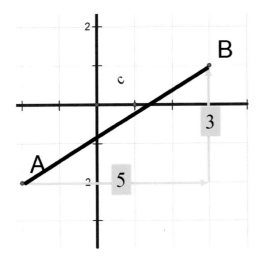

Q7) Use the Distance Formula to find the distance between points C (9, -4) and D (6, -9), to the nearest tenth.

Q8) What is the midpoint between points C and D?

Chapter 6

Algebra Review and Patterns

Section 6.1 Ratios and Proportions

What is a Ratio?

A ratio is a comparison of two numbers or terms. Ratios can be written in five different ways:

1) x to y
2) x: y
3) x ÷ y
4) $\frac{x}{y}$ ← Because a ratio is a fraction, y cannot be zero
5) x/y

How to Use Ratios?

The ratio of girls and boys is 15 **to** 10.

This means, for **every** 15 girls you can find 10 boys to match.
- There could be just 30 girls and 20 boys.
- There could be 45 girls and 30 boys.
- There could be 60 boys and 40 girls.

The ratio of 3 to 4 translates to $\frac{3}{4}$.

- The word "to" separates the numerator (15) and denominator (10) quantities.

Ratio can be expressed as in the following fraction: $\frac{x}{y}$, where y≠0.

Unit ratio: A ratio with a denominator of 1.

Example 1) Your school's soccer team has won 9 games and lost 2 games. What is the ratio of wins to losses?

<u>Solution:</u>

Because we are comparing wins to losses, the first number (9) in our ratio should be the number of wins and the second number (2) is the number of losses.

The ratio is $\dfrac{games \text{ won}}{games \text{ lost}} = \dfrac{9}{2}$.

Example 2) A bin at a hardware store contains 120 bolts and 150 nuts. Write the ratio of bolts to nuts in the simplest form.

<u>Solution:</u>

The ratio of **bolts to nuts** $= \dfrac{120}{150} = \dfrac{12}{15} = \dfrac{4}{5}.$

Example 3) The price of an 8 ounce can of tomato soup is $1.2. Write the unit ratio that expresses the price to weight.

<u>Solution:</u>

Weight = 8 oz.

Price = $1.2

The ration of price to weight $= \dfrac{1.2}{8} = 0.15$

The tomato soup costs $0.15 per ounce.

What is a Proportion?

A proportion is an equation that equates two ratios. In other words, an equation in which two ratios are equal is called a proportion.

Two equal ratios form a proportion.

Examples of equal ratios:

$\dfrac{5}{10} = \dfrac{1}{2}, \ \dfrac{6}{8} = \dfrac{3}{4}, \text{ and } \dfrac{3}{5} = \dfrac{6}{10}.$

Cross-Products of Proportions Property

For any algebraic expressions A, B, X, and Y. B & X do not equal zero then,

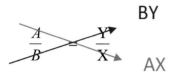

BY

AX

For example,

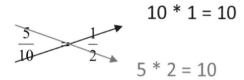

10 * 1 = 10

5 * 2 = 10

Thus,

$$\frac{5}{10} = \frac{1}{2} \Leftrightarrow 5*2 = 10*1.$$

Strategies for Solving Proportions

1. Calculate the cross products.
2. Set the cross products equal to each other.
3. Solve the equation (find the unknown value in the proportion).

Example 4) Solve the following proportion for x

$$\frac{7}{10} = \frac{x}{5}.$$

<u>Solution:</u>

$\frac{7}{10} = \frac{x}{5} \Leftrightarrow 7*5 = 10*x$ ⟵ Cross Product Property

$$35 = 10x$$

$$\frac{35}{10} = \frac{10x}{10}$$

$$3.5 = x$$

Example 5) Mary notices that her water bill was $30 for 500 cubic feet of water. At that rate, what would the charges be for 800 cubic feet of water?

Solution:

$$\frac{30}{500} = \frac{x}{800} \iff 30 * 800 = 500 * x$$ ← Cross Product Property

$$24000 = 500\ x$$

$$\frac{24000}{500} = \frac{500x}{500}$$

$$\$48 = x$$

Example 6) A car travels 160 miles in 3 hours. How far would it travel in 8 hours?

Solution:

$$\frac{160}{3} = \frac{x}{8} \iff 160 * 8 = 3 * x$$ ← Cross Product Property

$$1280 = 3\ x$$

$$\frac{1280}{3} = \frac{3x}{3}$$

$$426.7 \text{ miles} \approx x$$

Example 7) Three eighths of a garden is dug in 9 hours. How long will it take to dig to five eighths of it?

Solution:

$$\frac{3/8}{9} = \frac{5/8}{x} \iff \frac{3}{8} * x = \frac{5}{8} * 9$$ ← Cross Product Property

$$\frac{3x}{8} = \frac{45}{8}$$ ← Use Cross Product Property again

$$3x * 8 = 8*45$$

$$24\ x = 360$$

$$\frac{24x}{24} = \frac{360}{24}$$

$$x = 15 \text{ hours.}$$

Q1) How long does a 220-mile trip take moving at 50 miles per hour (mph)?

Q2) Two Tires cost $160. How much will four tires cost?

Q3) Solve the following Proportion:

$$\frac{2}{5} = \frac{x}{40}.$$

Q4) A classroom has 16 girls and 8 boys. What is the ratio of girls to boys in lowest terms?

Q5) The ratio of faculty members to students in one school is 1 to 25. There are 925 students. How many faculty members are there?

Q6) A house which is appraised for $100,000 pays $3000 in taxes. What should the tax be on a house appraised at $350,000?

Q7) It takes Shannon 3 hours to bike 12 miles. How long will it take her to bike 16 miles?

Time (hours)	Distance (miles)
3	12
X	16

Q8) Three sweaters cost $36. What is the cost of 5 sweaters?

Number of Sweaters	Cost
3	$36

Q9) Solve the following for x:

$$\frac{1}{8} = \frac{x}{5}.$$

Q10) $3\frac{1}{2}x = \frac{8}{9}$, then x =?

<u>**Definition**</u>

A number is written in scientific notation when it is expressed in the form:

$$n * 10^r$$

Where $1 \leq n < 10$ and r is an integer

Scientific Notation Rules

When expressing a number in scientific notation, remember the following rules:

Rule 1: Express the number as a **number** between 1 and 10 times a power of 10. In other words, to be in proper scientific notation the number must be written with:

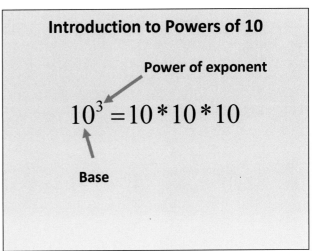

Introduction to Powers of 10

Power of exponent

$$10^3 = 10 * 10 * 10$$

Base

    a) a number between 1 and 10 ($1 \leq$ a number < 10)

    b) multiplied by a **power of ten**

The following are examples of scientific notations:

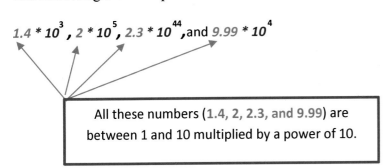

$1.4 * 10^3$, $2 * 10^5$, $2.3 * 10^{44}$, and $9.99 * 10^4$

All these numbers (1.4, 2, 2.3, and 9.99) are
between 1 and 10 multiplied by a power of 10.

And the following numbers are not scientific notations:

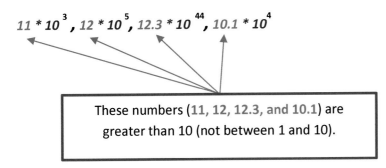

$11 * 10^3$, $12 * 10^5$, $12.3 * 10^{44}$, $10.1 * 10^4$

These numbers (11, 12, 12.3, and 10.1) are
greater than 10 (not between 1 and 10).

Rule 2: If the decimal point is moved to the left in the original number, make the power of 10 positive. If the decimal point is moved to the right in the original number, make the power of 10 negative.

Rule 3: The power of 10 always equals to the number of places that the decimal point has been shifted to the left or right in the original number.

To Write a number in Scientific Notation:

1) Move the decimal to the right of the first non-zero number

2) Count how many places the decimal had to be moved.

3) If the decimal had to be moved to the right, the exponent is negative.

4) If the decimal had to be moved to the left, the exponent is positive.

Note:

If you have an integer number, the decimal point will be located at the end of the number as in the following examples:

$5 = 5.0$

$77 = 77.0$

$123 = 123.0$

$699 = 699.0$

$7128 = 7128.0$

What is Standard Notation?

Standard notation is when a number is completely written out using numerical digits. Some examples of numbers written in standard notation are 54,150 and 8,000,000.

To Change from Standard Form to Scientific Notation:

1) Place the decimal point as there is one non-zero digit to the left of the decimal point.

2) Count the number of decimal places that the decimal has "moved" from the original number. This will be the exponent of the 10.

3) If the original number was less than 1, the exponent is negative; if the original number was greater than 1, the exponent is positive.

Example 1) Convert 749.3 from standard notation to scientific notation.

<u>Solution:</u>

1) The original number (749.3) is greater than 1 so the exponent is positive.
2) We are going to move the **decimal point** to the left between the first two digit numbers (7 & 4).

$$749.3 = 7.493 * 10^2$$

The exponent of 10 will be 2 because we moved the decimal point two places to the left.

Example 2) Convert 137,000,000 from standard notation to scientific notation.

<u>Solution:</u>

$$137,000,000 = 137,000,000.0$$

1) The original number (137,000,000) is greater than 1 so the exponent is positive.
2) We are going to move the **decimal point** to the left between the first two digit numbers (1 & 3).

$$137,000,000.0 = 1.370000000 * 10^8$$

The exponent of 10 will be 8 because we moved the decimal point 8 places to the left.

$$= 1.37 * 10^8$$

Example 3) Convert 7,490,000 from standard notation to scientific notation.

<u>Solution:</u>

$$7,490,000 = 7.49 * 10^6$$

The table below has some more examples of Standard Notations and their Scientific Notations:

Standard Form	Scientific Notation
123.4	$1.234 * 10^2$
2450	$2.45 * 10^3$
99000	$9.9 * 10^4$
897213.546	$8.97213546 * 10^5$

Example 4) In the United States, 15,000,000 households use private wells for their water supply. Write this number in scientific notation.

Solution:

$$15,000,000 = 15,000,000.0 = 1.5 * 10^7$$

Example 5) A ribosome (a cell part where protein is made), is about 0.000000003 of a meter in diameter. Write the length in scientific notation.

Solution:

1) The original number (0.000000003) is less than 1 so the exponent is negative.

2) We are going to move the decimal point to the right of number 3.

$$0.000000003 = 3 * 10^{-9}$$

The exponent of 10 will be **-9** because we moved the decimal point nine places to the right.

Mathematicians decided that by using powers of 10, they can create short versions of long numbers as in the following table:

Table : Powers of 10

$100,000,000 = 10^8$	$10 = 10^1$	$0.000001 = 10^{-6}$
$10,000,000 = 10^7$	$1 = 10^0$	$0.0000001 = 10^{-7}$
$1,000,000 = 10^6$	$0.1 = 10^{-1}$	$0.00000001 = 10^{-8}$
$100,000 = 10^5$	$0.01 = 10^{-2}$	$0.000000001 = 10^{-9}$
$10,000 = 10^4$	$0.001 = 10^{-3}$	$0.0000000001 = 10^{-10}$
$1,000 = 10^3$	$0.0001 = 10^{-4}$	$0.00000000001 = 10^{-11}$
$100 = 10^2$	$0.00001 = 10^{-5}$	$0.000000000001 = 10^{-12}$

Decimal Notation

A number written in standard form without using any form of powers of 10 notation is said to be written in decimal notation.

To Change from Scientific Notation to Standard Form:

1) If the exponent or power of 10 is positive, move the decimal point to the right, the same number of places as the exponent. In other words, if an exponent is positive, the number gets larger, so move the decimal to the right.

2) If the exponent or power of 10 is negative, move the decimal point to the left, the same number of places as the exponent. In other words, if an exponent is negative, the number gets smaller, so move the decimal to the left.

Example 6) Convert $1.85 * 10^4$ to standard notation.

Solution:

$1.85 * 10^4 = 18500$ ← We moved the decimal point 4 places to the right because the exponent of power 10 is positive (4).

Example 7) Convert $1.42 * 10^{-3}$ to decimal notation.

Solution:

$1.42 * 10^{-3} = 0.00142$ ← We moved the decimal point 3 places to the left because the exponent of power 10 is negative 3.

Example 8) The United States has a total of $1.2916 * 10^7$ acres of land reserved for state parks. Write this in standard form.

Solution:

$1.2916 * 10^7 = 12,916,000$ Acres

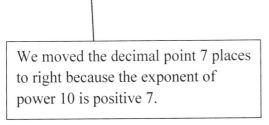

We moved the decimal point 7 places to right because the exponent of power 10 is positive 7.

Adding and Subtracting Numbers in Scientific Notation

When adding, or subtracting numbers in scientific notation, the exponents must be the same.

Steps to Add or Subtract Numbers with the Same Exponents:

1) Add or subtract the base numbers.
2) Bring down the given exponent on the 10.

So, when the exponents are the same for all the numbers you are working with, add or subtract the base numbers then simply put the given exponent on the 10 as in the following:

$$(A * 10^n) + (B * 10^n) = (A + B) * 10^n$$

$$(A * 10^n) - (B * 10^n) = (A - B) * 10^n$$

Where "A", "B" are the base numbers, and "**n**" is the exponent.

Example 9) Add the following scientific notations:

$$6.45 * 10^8 + 3.12 * 10^8$$

Solution:

$$(6.45 * 10^8) + (3.12 * 10^8) = (6.45 + 3.12) * 10^8$$

$$= 9.57 * 10^8$$

Example 10) Subtract the following scientific notations:

$$6.891 * 10^9 - 1.02 * 10^9$$

Solution:

$$(6.891 * 10^9) - (1.02 * 10^9) = (6.891 - 1.02) * 10^9$$

$$= 5.871 * 10^9$$

Steps to Add or Subtract Numbers with the Different Exponents:

1) To add or subtract numbers written in scientific notation, you must express them with the same power of ten.

2) Add or subtract the base numbers.

3) Bring down the given exponent on the 10.

Notes:

1) When adding, or subtracting numbers in scientific notation, the exponents must be the same.

2) If the powers of ten are different, you must move the decimal either right or left so that they will have the same exponent.

3) For each move of the decimal point to the **left** you should add $+1$ to the exponent as in the following examples:

$$53.1 * 10^4 = 5.31 * 10^{4+1} = 5.31 * 10^5$$

Here we move the decimal point one place to the left so we add **+1** to exponent.

$$423.02 * 10^9 = 4.2302 * 10^{9+2} = 4.2302 * 10^{11}$$

Here we move the decimal point two places to the left so we add **+2** to the exponent.

$$4283.11 * 10^6 = 4.28311 * 10^{6+3} = 4.28311 * 10^9$$

Here we move the decimal point three places to the left so we add **+3** to the exponent.

4) For each move of the decimal to the **right** you should add -1 to the exponent as in the following examples:

$$0.531 * 10^4 = 5.31 * 10^{4-1} = 5.31 * 10^3$$

Here we move the decimal point one place to the right so we add **-1** to the exponent.

$$0.023 * 10^9 = 2.3 * 10^{9-2} = 2.3 * 10^7$$

Here we move the decimal point two places to the right so we add **-2** to the exponent.

$$0.00411 * 10^6 = 4.11 * 10^{6-3} = 4.11 * 10^3$$

Here we move the decimal point three places to the left so we add **-3** to the exponent.

Example 11) Add the followings:

$(53.1 * 10^4) + (7.14 * 10^5).$

<u>Solution:</u>

$(53.1 * 10^4) + (7.14 * 10^5) =$

$(5.31 * 10^{4+1}) + (7.14 * 10^5) =$

$(5.31 * 10^5) + (7.14 * 10^5) =$

$(5.31 + 7.14) * 10^5 =$

$12.45 * 10^5$

Example 12) Add the followings:

$(0.0566 * 10^8) - (2.55 * 10^6)$.

<u>Solution:</u>

$(0.0566 * 10^8) - (2.55 * 10^6) =$

$(5.66 * 10^{8-2}) - (2.55 * 10^6) =$

$(5.66 * 10^6) - (2.55 * 10^6) =$

$(5.66 - 2.55) * 10^6 =$

$3.11 * 10^6$

Exercise Set 6.2

Q1) Convert 1349.7 from standard notation to scientific notation.

Q2) Convert $546.34 * 10^{-4}$ to standard notation.

Q3) Write 0.00705 in scientific notation.

Q4) A certain cell has a diameter of approximately $4.7065 * 10^{-6}$ meters. A second cell has a diameter of $4.69 * 10^{-6}$ meters. Which cell has a greater diameter?

Q5) Subtract $2.0 * 10^7$ from $8.0 * 10^7$

Q6) $(4.12 * 10^6) + (3.94 * 10^4) =$?

Q7) $(4.23 \times 10^3) - (9.56 \times 10^2) =$?

Q8) Convert the following scientific number into decimal notation

$7.47 * 10^{-3}$

Q9) $(100,000,000)^0 =$?

Q10) $10^{-6} =$?

Section 6.3 Multiplying and Dividing Numbers in Scientific Notation

Rules of Multiplying and Dividing Numbers in Scientific Notation

Rule#1: To multiply numbers written in scientific notation, do the following:

    1) Multiply the coefficients (base numbers) together.
    2) Add the exponents.
    3) The base will remain 10.
    4) Put your answer in scientific notation.

Rule#2: To divide numbers written in scientific notation, do the following:

    1) Divide the coefficients (base numbers).
    2) Subtract the exponents.
    3) The base will remain 10.
    4) Put your answer in scientific notation.

The general formats for multiplying and dividing numbers in scientific notation are as follows:

$$(A * 10^x) * (B * 10^y) = (A * B) * 10^{x+y}$$

and

$$\frac{(A * 10^x)}{(B * 10^y)} = \left(\frac{A}{B}\right) * 10^{x-y} .$$

Product of Powers

To multiply powers with the same base, add their exponents.

For example, $10^x * 10^y = 10^{x+y}$.

Where "A" and "B" are the coefficients (base numbers), and "**x**" and "**y**" are the exponents.

Example 1) Multiply the following then write your answer in scientific notation:

$$(4 * 10^2) * (8 * 10^5).$$

<u>Solution:</u>

$$(4 * 10^2) * (8 * 10^5) = (4 * 8) * 10^{2+5} = 32 * 10^7 = 3.2 * 10^{7+1} = 3.2 * 10^8 .$$

Example 2) Multiply the following:

$(-1.8 * 10^7) * (-2.9 * 10^{-2})$.

Solution:

$(-1.8 * 10^7) * (-2.9 * 10^{-2}) = (-1.8 * -2.9) * 10^{(7) + (-2)} = +5.22 * 10^5 = 5.22 * 10^5$.

Example 3) Divide the following:

$$\frac{(8 * 10^9)}{(2 * 10^2)}.$$

Solution:

$$\frac{(8 * 10^9)}{(2 * 10^2)} = \left(\frac{8}{2}\right) * 10^{9-2}$$

$$= 4 * 10^7.$$

> **Quotient of Powers**
>
> To divide powers with the same base, subtract their exponents.
>
> For example, $10^x \div 10^y = 10^{x-y}$

Example 4) Divide the following:

$$\frac{(4.86 * 10^9)}{(2.0 * 10^3)}.$$

Solution:

$$\frac{(4.86 * 10^9)}{(2.0 * 10^3)} = \left(\frac{4.86}{2}\right) * 10^{9-3}$$

$$= 2.43 * 10^6.$$

Q1) Multiply the following then write your answer in scientific notation:

$(9 * 10^2) * (5 * 10^4)$.

Q2) $\left[9 * 10^5 \right]^2 = ?$

Q3) Divide the following:

$$\frac{(16 * 10^9)}{(2 * 10^2)} \; .$$

Q4) Multiply the following then write your answer in scientific notation:

$(9.125 * 10^3) * (5.04 * 10^{25})$.

Q5) Multiply $(3.2 * 10^{-3}) (2.1 * 10^5)$.

Q6) Divide $(6.4 * 10^6)$ by $(1.7 * 10^2)$.

Q7) Evaluate 3,600,000,000 * 24 (write your answer in scientific notation).

Q8) $(9.12 * 10^5) * (2.35 * 10^7) = ?$

Q9) $(8 * 10^6) \div (2 * 10^3) = ?$

Q10) A rectangular section of land made up of cotton farms has a length of 8.3 * 105 meters and a width

7.7 * 104 meters. What is the area of the land in square meters?

A set of equations is called a **system of equations**. The solutions must satisfy each equation in the system. If all equations in a system are linear, the system is **a system of linear equations**, or a **linear system.**

Systems of Linear Equations:

A solution to a system of equations is an ordered pair that satisfy all the equations in the system.

There are **4 methods** of solving a system of equations algebraically:

1) By graphing.

2) By substitution.

3) By elimination (also called addition).

4) By multiplication.

1) Solving Linear Systems by Graphing

Strategy for Solving Linear Systems by Graphing

1) Find ordered pairs that satisfy each of the equations.
2) Plot the ordered pairs and sketch the graphs (it should be a straight line) of both equations on the same axis.
3) The coordinates of the point or points of intersection of the graphs are the solution or solutions to the system of equations.

There are three possible solutions to a linear system in two variables:

1) One solution: Coordinates of a point (consistent).
2) No solution: Inconsistent case.
3) Infinite number of solution: Dependent case.

Below is a table that represents number of solutions of a linear system:

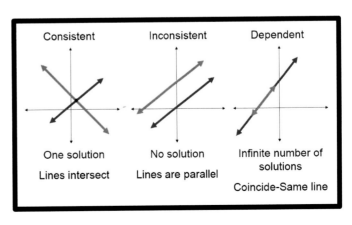

175

Example 1) Solve the following system by graphing:

$$x - y = 2$$

$$y = -\frac{2}{3}x + 3.$$

<u>Solution:</u>

We need to find two ordered pairs that satisfy the first equation $(x - y = 2)$.

1) Choose any number for x such as $x = 0$ then find y in the first equation below:

$$x - y = 2$$
$$0 - y = 2$$
$$y = -2.$$

The first order pair for equation $x - y = 2$ is $(0, -2)$.

2) Choose any number for y such as $y = 0$ then find x in the first equation below:

$$x - y = 2$$

$$x - 0 = 2$$

$$x = 2.$$

The second order pair for equation $x - y = 2$ is $(2, 0)$.

Draw the first equation by plotting both order pairs $(0, -2)$ and $(2, 0)$ then connect the two points with a straight line as below:

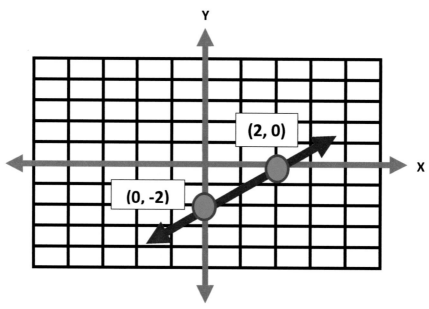

We also need to find two ordered pairs that satisfy the second equation ($y = -\frac{2}{3}x + 3$).

1) Choose any number for x such as x = 0 then find y in the second equation below:

$$y = -\frac{2}{3}x + 3$$

$$y = -\frac{2}{3}(0) + 3$$

$$y = 3.$$

The first order pair for equation $y = -\frac{2}{3}x + 3$ is (0, 3).

2) Choose any number for y such as y = 0 then find x in the second equation below:

$$y = -\frac{2}{3}x + 3$$

$$0 = -\frac{2}{3}x + 3 \implies \frac{2}{3}x = 3 \implies \frac{2x}{3} = 3 \implies 2x = 9 \implies x = \frac{9}{2} = 4.5$$

The second order pair for equation $y = -\frac{2}{3}x + 3$ is (4.5, 0).

Draw the second equation by plotting both order pairs (0, 3) and (4.5, 0) then connect the two points with a straight line as below:

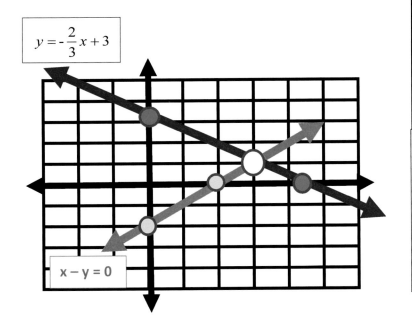

From the graph, the intersection point between the two lines is **(3, 1)** which is the solution to the system:

$$x - y = 2$$

$$y = -\frac{2}{3}x + 3$$

NOTES

1) This system of equations represents two intersecting lines.
2) This system of equations has one solution.

2) Solving Linear Systems by Substitution

Strategy for Solving Linear Systems by Substitution

1) Solve one equation for one of the variables in terms of the other. (If one of the equations is already in this form, you can skip this step). In other words, choose one of the two equations and isolate one of the variables.
2) Substitute the expression found in step 1 into the other equation. This will result in an equation in one variable. In other words, substitute the value of the variable into the other equation.
3) Simplify and solve the equation containing one variable.
4) Back-substitute the value found in step 3 into the equation from step 1. Simplify and find the value of the remaining variable. In other words, substitute back into either equation to find the value of the other variable.
5) Check the proposed solution in both of the system's given equations.

Example 2) Solve the system below using substitution method:

$$y = 2x - 8$$

$$2x + 3y = 16.$$

Solution:

Step 1: Solve either of the equations for one variable in terms of the other.

Notice that the first equation is already solved for y, so we will use that.

Solve using substitution method:
$$y = 2x - 8$$
$$2x + 3y = 16.$$

Step 2: Substitute the expression from step 1 into the other equation.

$$2x + 3y = 16$$
$$2x + 3(2x - 8) = 16$$

Substitute $2x - 8$ for y

Step 3: Solve the resulting equation.

$$2x + 3(2x - 8) = 16$$
$$2x + 6x - 24 = 16$$
$$8x - 24 = 16$$
$$8x = 40$$
$$x = 5.$$

Step 4: Back-substitute the obtained value into the equation from step 1.

Substitute $x = 5$ into the first equation below:

$y = 2x - 8$

$y = 2(5) - 8$

$y = 10 - 8$

$y = 2.$

The proposed solution is $x = 5$ and $y = 2$ or the ordered pair $(5, 2)$.

Step 5: Check the proposed solution $(5, 2)$ in both equations.

Equation 1	**Equation 2**
$y = 2x - 8$	$2x + 3y = 16$
$2 = 2(5) - 8$	$2(5) + 3(2) = 16$
$2 = 10 - 8$	$10 + 6 = 16$
$2 = 2$ True.	$16 = 16$ True.

Therefore, $(5, 2)$ satisfies both equations. The system's solution is $(5, 2)$.

3) Solving Linear Systems by Elimination (Addition)

Strategy for Solving Linear Systems by Elimination

1) Write both equations in Standard Form, $Ax + By = C$.

2) Determine which variable to eliminate with Subtraction or Addition. The key for solving a system by elimination is getting rid of one variable.

3) Solve for the variable left.

4) Go back and use the found variable in step 3 to find second variable.

5) Check the solution in both equations of the system.

Example 3) Solve the following system using the elimination method:

$$5x + 3y = 11$$

$$5x = 2y + 1$$

Solution:

1) Write both equations in $Ax + By = C$

$$5x + 3y = 11 \qquad\qquad\qquad 5x + 3y = 11$$

$$5x = 2y + 1 \qquad\longrightarrow\qquad 5x - 2y = + 1$$

2) Determine which variable to eliminate with Addition or Subtraction. Use subtraction to eliminate 5x.

$$5x + 3y = 11 \qquad\qquad\qquad 5x + 3y = 11$$

$$- \left[(5x - 2y) = + 1 \right] \qquad\longrightarrow\qquad -5x + 2y = -1$$

Here we change the sign of each term in the 2nd equation to its opposite sign because we are going to subtract equation 2 from equation 1. We eliminate 5x and solve for y.

$$5x + 3y = 11$$

$$-5x + 2y = -1$$

$$\overline{}$$

$$5y = 10$$

$$\frac{5y}{5} = \frac{10}{5}$$

$$y = 2.$$

3) Solve for the other variable (x) by substituting $y = 2$ into either equation.

$$5x + 3y = 11$$

$$5x + 3(2) = 11$$

$$5x + 6 = 11$$

$$5x = 5$$

$$x = 1.$$

The solution to the system is $(1, 2)$.

4) Check the solution in both equations. The solution to the system is $(1, 2)$.

$5x + 3y = 11$

$5(1) + 3(2) = 11$

$5 + 6 = 11$

$11 = 11$ True.

$5x = 2y + 1$

$5(1) = 2(2) + 1$

$5 = 4 + 1$

$5 = 5$ True.

Example 4) The sum of 5 times a first number and second number is 23 and the difference of 5 times the first number and second number is 17. Find these two numbers.

<u>Solution:</u>

Let x = first number and

y = second number.

Therefore,

$5x + y = 23$

$5x - y = 17$

1) Both equations are already in standard form $Ax + By = C$

$5x + y = 23$

$5x - y = 17$

2) Add both equations to eliminate variable y.

$5x + y = 23$

$5x - y = 17$

$10x = 40$

$\dfrac{10x}{10} = \dfrac{40}{10}$

$x = 4$ (this is the first number).

3) Solve for the other variable (y) by substituting $x = 4$ into either equation.

$$5x + y = 23$$

$$5(4) + y = 23$$

$$20 + y = 23$$

$$y = 23 - 20$$

$$y = 3 \text{ (this is the second number)}.$$

4) Check the solution in both equations. The solution to the system is $(4, 3)$.

$5x + y = 23$

$5(4) + 3 = 23$

$20 + 3 = 23$

$23 = 23$ True.

$5x - y = 17$

$5(4) - 3 = 17$

$20 - 3 = 17$

$17 = 17$ True.

---

4) Solving Linear Systems by Multiplication

Strategy for Solving Linear Systems by Multiplication

1) Write both equations in Standard Form, $Ax + By = C$. If none of the coefficients are 1 or -1 and neither of the variables can be eliminated by simply adding and subtracting the equations, multiply one or both equations by a number so that the coefficients of a chosen variable are opposites.

2) Eliminate the chosen variable with subtraction or addition.

3) Solve for the variable left.

4) Go back and use the found variable in step 3 to find second variable.

5) Check the solution in both equations of the system.

Example 5) Solve the following system using the elimination method:

$$10x + 3y = 41$$

$$-5x + 2y = 4$$

<u>Solution:</u>

1) Both equations are already in standard form $Ax + By = C$.

$$10x + 3y = 41$$

$$-5x + 2y = 4$$

2) Multiply the second equation by 2.

$$10x + 3y = 41$$

$$2 \left[-5x + 2y = 4 \right] \qquad\longrightarrow\qquad$$

$$10x + 3y = 41$$

$$-10x + 4y = 8$$

3) Add the new equation

$$10x + 3y = 41$$

$$\underline{-10x + 4y = 8}$$

$$7y = 49$$

$$\frac{7y}{7} = \frac{49}{7}$$

$$y = 7.$$

4) Solve for the other variable (x) by substituting $y = 7$ into either equation.

$$10x + 3y = 41$$

$$10x + 3(7) = 41$$

$$10x + 21 = 41$$

$$10x = 41 - 21$$

$$10x = 20$$

$$x = 2.$$

The solution to the system is $(2, 7)$.

5) Check the solution in both equations. The solution to the system is $(2, 7)$.

$$10x + 3y = 41$$
$$10(2) + 3(7) = 41$$
$$20 + 21 = 41$$
$$41 = 41 \text{ True.}$$

$$-5x + 2y = 4$$
$$-5(2) + 2(7) = 4$$
$$-10 + 14 = 4$$
$$4 = 4 \text{ True.}$$

Exercise Set 6.4

Q1) Solve the following system by using substitution method:

$$2x - y = 2$$
$$y = -2 - x$$

Q2) Solve the following system by graphing:

$$x - y = -1$$
$$x + 2y = 5$$

Q3) Solve the following system by using elimination method:

$$x - y = 1$$
$$x + y = 9$$

Q4) Find two numbers whose sum is 18 and whose difference is 22.

Q5) One number is 4 more than twice the second number. Their total is 25. Find the numbers.

Q6) Graph the following system to find the solution:

$x + y = -2$

$2x - 3y = -9$

Q7) Kim is 3 years older than Jim. In five years the sum of their ages will be 68. How old are they now?

Q8) The sum of two numbers is 120 and their difference is 10. Find the numbers.

Q9) The sum of two numbers is 81. One number is 41 more than the other. Find the numbers using the substitution method.

Q10) Solve the following system of linear equation by using addition method.

$2x - y = 4$

$4x + 2y = 16$

Q11) What is the solution of the system of equations below?

$x + 2y = 7a$

$3x - 2y = 5a$

a. (3a, 2a)

b. (-3a, 2a)

c. (-3a, -2a)

d. (4, 4)

e. (2, 0)

Section 6.5 Rate, Distance, and Time

Rate

Rate is the speed of motion, change, or activity.

A **physical distance** can be meant several different things:

➤ The length of a specific path traveled between two points, such as the distanced walked while navigating a maze.
➤ The length of the shortest possible path through space between two points that could be taken if there were no obstacles.
➤ The length of the shortest path between two points such as the shortest distance between any two places.

Calculation Rate, Distance, and Time

The rate formula is given as:

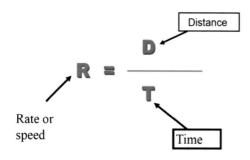

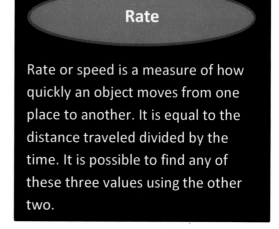

Rate

Rate or speed is a measure of how quickly an object moves from one place to another. It is equal to the distance traveled divided by the time. It is possible to find any of these three values using the other two.

The distance formula is given by:

D = R * T where;

D is the distance

R is the rate (speed)

T is the time.

Measurement

Measurement is defined as the comparison between unknown quantities with known quantities of same kind. To approach the subject quantitatively, it is essential that we make measurements. The following table lists some units in common use for speed and their abbreviations:

Distance	Time	Rate (Speed)	Abbreviation
mile	hours	miles per hour	mph
kilometers	hours	kilometers per hour	km/h
meters	hours	meters per hour	m/h
meters	seconds	meters per second	m/second or m/s
centimeters	seconds	centimeters per second	cm/s
feet	seconds	feet per second	ft. per sec. or f.p.s.
inches	seconds	inches per second	in/sec or in/s

Conversion Key Points

1) Time Conversion

❖ To convert minutes to hours, divide by 60.

❖ To convert hours to minutes, multiply by 60.

❖ To convert seconds to minutes, divide by 60.

❖ To convert minutes to seconds, multiply by 60.

❖ To convert days to hours, multiply by 24.

❖ To convert hours to days, divide by 24.

2) Measurement Conversions

❖ To convert inches to feet, divide by 12.

❖ To convert feet to inches, multiply by 12.

❖ To convert feet to miles, divide by 5280.

❖ To convert miles to feet, multiply by 5280.

❖ To convert kilometers to meters, multiply by 1000.

❖ To convert meters to kilometers, divide by 1000.

❖ To convert meters to centimeters, multiply by 100.

❖ To convert centimeters to meters, divide by 100.

Example 1) A train travels 457 miles between Boston and Washington, the trip takes 6.5 hours. What is the average speed?

Solution:

D = 457 miles

T = 6.5 hours

R=?

Distance (miles)	=	Rate (miles/hour)	·	Time (hours)
⬇		⬇		⬇
457	=	R	·	6.5

457 = 6.5 R

$$\frac{457}{6.5} = \frac{6.5R}{6.5}$$

70.3 = R

R = 70.3 miles/hour.

The average speed of the train is about 70.3 miles per hour.

Example 2) A Delta airplane flies at an average speed of 540 miles per hour. How long will it take to fly from New York to Tokyo knowing that the distance between the two cities is 6760 miles?

Solution:

R = 540 miles
T = ?
D = 6760 miles

D = R * T

6760 = 540T

T= 12.5 hrs.

Example 3) The distance to the moon is $2.36 * 10^5$ miles. How long does it take a rocket traveling at 3.2×10^3 mph to reach the moon?

Solution:

$d = 2.36 * 10^5$ miles

$r = 3.2 * 10^3$ miles/hr

$t = ?$

$$t = \frac{d}{r} = \frac{2.36 * 10^5}{3.2 * 10^3} = \frac{2.36}{3.2} * \frac{10^5}{10^3}$$

$t = 0.7375 * 10^2 = 73.75 \ hrs.$

Example 4) How far did an airplane go from Atlanta to Istanbul if it travels at 880 km/hr. for 10 hours and 30 minutes?

Solution:

D =? km

T = 10 hours and 30 minutes

R= 880 km/hr.

30 minutes = 30/60 = 0.5 hr.

T = 10.5 hrs.

Distance (km)	=	Rate (km/hour)	·	Time (hours)
↓		↓		↓
D	=	880	·	10.5

D = 9,240 km.

Q1) How long will a train take to travel 667.5km at 89km/hr.?

Q2) A runner's average speed is 7m/s, how far does she run in 30 seconds?

Q3) Car travels 50km in 30 minutes. What is its average speed in km/h?

Q4) How far can you go in 20 minutes at 55 miles per hour?

Q5) Find the speed of a train in mph if it takes 5 hours to travel a distance of 180 miles.

Q6) A salesman travelled at an average speed of 50 km/h for 2 hours 30 minutes. How far did he travel?

Q7) A train travelled 555 miles at an average speed of 60 mph. How long did the journey take?

Q8) You ride your bike for 3 hours. If you travel 36.75 miles, what is your average speed?

Q9) Hari can type 900 words in 20 minutes. Calculate his typing speed in:

    a) Words per minute.

    b) Words per hour.

Q10) Heleen drives 72 miles from Rome, GA to Chattanooga, TN in one hour and 15 minutes. She then drives 135 miles from Chattanooga, to Nashville, TN, in 2 hours. Determine her average rate for each journey.

Section 6.6 Simple Interest

Interest

Interest is money paid for the use of money. If you borrow from the bank to buy a car, the bank will charge you interest for its use. If you open a saving account at the bank, the bank will pay you interest if the account is open.

Simple Interest

Simple interest is interest that is computed on the original sum. The formula for calculating simple interest is

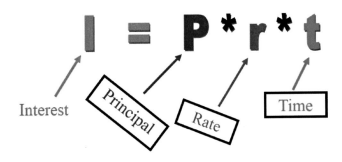

Hint

Remember, when symbols are written side by side, it means to multiply, so *Prt* means *P* * *r* * *t*. Also, don't forget *r*, the interest rate, is the *annual* rate; and *t* is the time expressed in *years* (or a fraction of a year).

Where:

I	P	r	t
Interest (amount earned or paid)	principal (the amount invested or borrowed)	rate of interest (yearly usually written as decimal)	time of loan or investment (usually in one year or years)

Example 1) You want to borrow $4000 at 12% for only 90 days. How much interest will you pay?

Solution:

I =?

P = $4000

r =12% = 0.12

t= 90 days= 3 months= 3/12= 0.25 Year

I =   P * r * t

I = 4000 * 0.12 * 0.25 = $120.

Remember to express 3 months as 3/12 year in the formula.

Example 2) Shelby needs to borrow $190,000 to buy a house. She can borrow the money for 20 years at 4.3%. How much interest will she pay for the loan?

Solution:

I =?

P= $190,000

r = 4.3%= 0.043

t = 20 years

Shelby will pay the interest plus the amount she borrowed.

I =   P * r * t

I = 190,000*0.043*20

I = $163,400.

Example 3) Mary borrows $25,000 to buy a car at 2.7% interest.

   1) How much interest will she pay if the loan is for 5 years?

   2) How much total amount will Mary pay for the car?

   3) How much will Mary have to pay each month to the bank for the car?

<u>Solution:</u>

  1) I =?

P= $25,000

r = 2.7%= 0.027

t = 5 years

I = P * r * t

I = 25,000 * 0.027 * 5 = $3, 375 interest.

> **Fit More Steps into Your Day!**
>
> ✓ Take several 10 minutes' walks during the day.
> ✓ Take the stairs instead of the elevator.
> ✓ Pace while talking on the phone.
> ✓ Make family walks a habit.

  2) How much total amount will Mary pay for the car?

25,000 + 3,375 = $28,375.

  3) How much will Mary have to pay each month to the bank for the car?

5 years * 12 months = 60 months

28,375 ÷ 60 = $472.91

Hint

    Some students like to place the symbols I, p, r, and t in a circle (notice that the I is alone at the top). The equation for each of the variables is found by covering the appropriate letter as in the following:

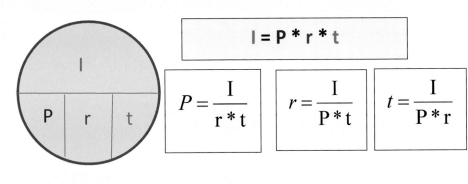

$$I = P * r * t$$

$$P = \frac{I}{r * t} \qquad r = \frac{I}{P * t} \qquad t = \frac{I}{P * r}$$

Q1) Melissa saved her $20,000 for 3 ½ years at 2.15% APR in a CD, to go on a month-long vacation with her family. How much did she earn in interest?

Q2) Judy bought a new motorcycle for $28,000. She took out a loan for 4 ½ years with 1.75% annual percentage rate (APR). How much will Judy end up paying in interest?

Q3) If you get a student loan to pay for your educational expenses this year. Find the interest on the loan if you borrowed $2,000 at 8% for 1 year.

Q4) Jordan has $45,560 in a savings account that pays 0.15% simple interest. How much interest will she earn in each amount of time?

    a) 5 years   **b)** 6 months

Q5) Darren bought four new tires for $900 using a credit card. His card has an interest rate of 19%. If he has no other charges on his card and does not pay off his balance at the end of the month, how much money will he owe after one month?

Q6) An office manager charged $625 worth of office supplies on a charge card with an interest rate of 14% APR. How much money will he owe if he makes no other charges on the card and does not pay off the balance at the end of the month?

Q7) Find the rate of simple interest charged per APR if a loan of $20,000 incurs interest of $12,000 after eight years.

Q8) $8,000 is invested for 1 year at a rate of 3%. How much interest is earned? What is the total amount after one year?

Q9) Find the length of time it would take for $50,000 invested at an interest rate of 8% to earn $10,000 interest.

Q10) Find the amount that should be invested to earn $1,350 interest over 3 years at an interest rate of 4.5% APR.

The Square Root Property

If $x^2 = a$, where "a" is a real number, then $x = \pm \sqrt{a}$. The square root property is one method that is used to find the solutions to a quadratic (second degree) equation. This method involves taking the square roots of both sides of the equation as in the following examples:

$$\text{if } x^2 = 5 \text{ then } x = \pm\sqrt{5},$$

$$\text{if } x^2 = 36 \text{ then } x = \pm\sqrt{36} = \pm 6, \text{ and}$$

$$\text{if } (x+y)^2 = b \text{ then } (x+y) = \pm\sqrt{b}.$$

Example 1) Solve the following quadratic equation:

$$4x^2 = 81.$$

<u>Solution:</u>

Before taking the square root of each side, you must isolate the term that contains the squared variable.

$4x^2 = 81$ | This is the given equation.

$\dfrac{4x^2}{4} = \dfrac{81}{4}$ | Divide both sides by 4 to isolate the x.

$x^2 = \dfrac{81}{4}$ | Apply the square root property.

$x = \pm\sqrt{\dfrac{81}{4}}$

$x = \pm\dfrac{\sqrt{81}}{\sqrt{4}}$ | Simplify.

$x = \pm\dfrac{9}{2}.$

195

Example 2) Solve the following equation:

$$x^2 - 16 = 0.$$

by using:

    1) The square root property.
    2) The factoring method.

<u>Solution:</u>

    1) Using the square root property

$$x^2 - 16 = 0$$

| This is the given equation. |

$$x^2 = 16$$

| Isolate x^2 in one side (left side). |

$$x = \pm\sqrt{16}$$

| Apply the square root property. |

$$x = \pm 4$$

| Simplify. |

    2) Using Factoring method (the difference between two squares). Review section 4.4 (Page 118).

$$x^2 - 16 = 0$$

$$x^2 - 4^2 = 0$$

$$(x - 4)(x + 4) = 0$$

$$(x - 4) = 0 \text{ or } (x + 4) = 0$$

$$x = +4 \quad \text{or} \quad x = -4$$

Notice that:

    5) $x^2 - 16$ is a binomial.

    6) The first term is x^2 which is a perfect square.

    7) The last term is **16** which is a perfect square because $16 = 4^2$.

    8) There is a subtraction between x^2 and **16.**

Example 3) Solve $(x+2)^2 = 9$.

Solution:

$(x+2)^2 = 9$

$x+2 = \pm\sqrt{9}$

$x+2 = \pm 3$ (Work it twice $\pm$).

$x = -2 + 3$ or $x = -2 - 3$

$x = 1$ or $x = -5$

$x = \{-5, 1\}$.

Example 4) Solve the following quadratic equation by using the square root property:

$$6x^2 - 0.06 = 0$$

Solution:

$6x^2 - 0.06 = 0$

$6x^2 = 0.06$

$\dfrac{6x^2}{6} = \dfrac{0.06}{6}$

$x^2 = 0.01$

$x = \pm\sqrt{0.01}$

$x = \pm 0.1$

x = $\{-0.1, +0.1\}$.

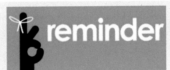
reminder

➤ The square root property states that if we take the square root of each side of an equation, the result is an equivalent equation.
➤ The squared quantity in a quadratic equation must be **isolated** before the square root property can be applied.
➤ The square root property results in **two** new equations! One is positive and the other one is negative.

Q1) Solve the following quadratic equation by using the square root property:

$$4x^2 = 28.$$

Q2) Solve the following quadratic equation by using the square root property:

$$3(x+2)^2 = 36.$$

Q3) If a square bedroom has an area of 121 square feet, what is the length of one wall?

Q4) Solve $3x^2 - 48 = 0$.

Q5) A square garden has an area of 144 square feet. How much fencing will a gardener need to buy in order to place fencing around the garden?

Q6) The area of a rectangle is 20 square meters. Find the length and width if the length is 1.5 times the width.

Q7) Solve $x^2 + 5 = 167$.

Q8) The product of two positive numbers is 140. Determine the numbers if the larger is 5.6 times the smaller.

Q9) Solve the equation: $x^2 - 25 = 0$.

Q10) Solve $(3x - 4)^2 = 42$.

H
O
M
E
W
O
R
K

Section 6.8 Absolute Value Equations and Inequalities

Absolute Value Equations

When you take the absolute value of a number, you always end up with a positive number (or zero). Whether the input was positive or negative (or zero), the output is always positive (or zero).

For instance, $|5| = 5$, and $|-5| = 5$. See the absolute value on Page 17 for more details.

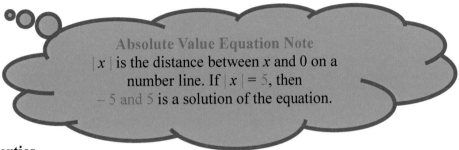

Absolute Value Equation Note
$|x|$ is the distance between x and 0 on a number line. If $|x| = 5$, then -5 and 5 is a solution of the equation.

Absolute Value Equations Properties

Property #1) If **x** is a positive real number, **a** represents any algebraic expression, and

$|x| = a$ then $x = a$ or $x = -a$. See Table#1 below.

Property #2) If **x** is a positive real number, **y** represents any algebraic expression, and

$|x| = |y|$ then $x = y$ or $x = -y$.

Notes:

1) An equation containing a variable within an absolute value symbol is called an absolute value equation.

2) Absolute values can never be negative in value.

3) Distance is always a positive number or zero. Therefore, the absolute value of a number is always a positive number or zero.

Table# 1: Absolute Value Equations		
Absolute Value Equation	**Equivalent Equation**	**Solution Set**
$\lvert x \rvert = a \ (a > 0)$	x = a or x = -a	{a, -a}
$\lvert x \rvert = 0$	x = 0	{0}
$\lvert x \rvert = a \ (a < 0)$		∅

Strategy to Solve Absolute Value Equations:

1) Isolate the absolute value (let the absolute value by itself on one side of an equation).

2) Solve the equation using Property#1.

3) Check the answer.

Example 1) Solve $|2a - 1| - 3 = 4$.

<u>Solution:</u>

$|2a - 1| - 3 = 4$ Isolate the absolute value by adding +3 to both sides of the equation.

$\qquad +3 \quad +3$

$|2a - 1| = 4 + 3$

$|2a - 1| = 7.$ Solve the equation using Property#1.

$2a - 1 = +7 \quad$ or $\quad 2a - 1 = -7$

$2a = +7 + 1 \qquad\qquad 2a = -7 + 1$

$2a = 8 \qquad\qquad\qquad 2a = -6$

$a = 4. \qquad\qquad\qquad a = -3.$

Check your solution

When a = 4	When a = -3
$|2a - 1| - 3 = 4$	$|2a - 1| - 3 = 4$
$|2(4) - 1| - 3 = 4$	$|2(-3) - 1| - 3 = 4$
$|8 - 1| - 3 = 4$	$|-6 - 1| - 3 = 4$
$|7| - 3 = 4$	$|-7| - 3 = 4$
$7 - 3 = 4$ True.	$7 - 3 = 4$ True.

Example 2) Solve the following equation:

$| \frac{2}{3}x - 3 | + 2 = 12.$

Solution:

$| \frac{2}{3}x - 3 | + 2 = 12$ | Isolate the absolute value by adding –2 to both side of the equation. |

$| \frac{2}{3}x - 3 | = 12 - 2$

$| \frac{2}{3}x - 3 | = 10$ | Solve the equation using Property#1. |

$\frac{2}{3}x - 3 = +10$ or $\frac{2}{3}x - 3 = -10$

$\frac{2}{3}x = +10 + 3$ $\frac{2}{3}x = -10 + 3$

$\frac{2}{3}x = +13$ $\frac{2}{3}x = -7$

$\frac{3}{2}\left(\frac{2}{3}x \right) = +13\left(\frac{3}{2} \right)$ $\frac{3}{2}\left(\frac{2}{3}x \right) = -7\left(\frac{3}{2} \right)$

$x = +\frac{39}{2}.$ $x = -\frac{21}{2}.$

Example 3) Solve the following equation:

$|2x - 3| = |x - 9|$.

<u>Solution:</u>

$|2x - 3| = |x - 9|$ | Solve the equation using Property#2. |

$2x - 3 = x - 9$ or $2x - 3 = -(x - 9)$

$2x - x = -9 + 3$ | $2x - 3 = -x + 9$

$x = -6.$ | $2x + x = +9 + 3$

| $3x = 12$

| $x = 4.$

Absolute Value Inequalities

Absolute value inequalities have infinitely many solutions. We call this a solution set.

Properties of Absolute Value Inequalities

The following properties of the absolute value function need to be memorized:

Property#1) For any real number **x**, and any nonnegative number **a**, and | **x** | < **a** then

-**a** < **x** < a. This property is also valid if < is replaced by ≤.

For example, if $| x | < 5$, then any number between -5 and 5 is a solution of this inequality.

Property#2) For any real number **x**, and any nonnegative number **a**, and | **x** | > **a** then

x > **a** or **x** < -**a**. This property is also valid if > is replaced by ≥. See Table#2 below:

Table# 1: Absolute Value Inequalities		
Absolute Value Inequality	**Equivalent Inequality**	**Solution Set**
$|x| < a$	-a < x < a	(a, -a)
$|x| \leq a$	-a ≤ x ≤ a	[-a, a]
$|x| > a$	x > a or x < -a	(-∞, -a) ∪ (a, ∞)
$|x| \geq a$	x ≥ a or x ≤ -a	(-∞, -a] ∪ [a, ∞)

Example 4) Solve the following inequality and graph the solution set on a number line:

$|x - 2| < 5.$

Solution:

$|x - 2| < 5$ | Solve the equation using Property#1 (rewrite without absolute value bars). |

$$-5 < x - 2 < 5$$
$$-5 + 2 < x - 2 + 2 < 5 + 2$$
$$-3 < x < 7$$

The solution set is $(-3, 7)$.

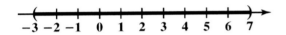

Example 5) Solve the following inequality:

$|x| > 8.$

Solution:

$|x| > 8$ | Solve the equation using Property#2 (rewrite without absolute value bars). |

$x > 8$ or $x < -8$

Example 6) Solve the following inequality:

$|x-1| \geq 8.$

Solution:

$|x-1| \geq 8$ | Solve the equation using Property#2 (rewrite without absolute value bars). |

$x-1 \geq 8$ or $x-1 \leq -8$

$x \geq 8+1$ or $x \leq -8 +1$

$x \geq 9.$ or $x \leq -7.$

Exercise Set 6.8

Q1) Solve $|2x - 1| = 7$.

Q2) Solve $|y - 4| = \dfrac{5}{3}$.

Q3) Solve: $3|a + 5| = 12$.

Q4) Solve $|\dfrac{2}{3}a - 3| + 5 = 12$.

Q5) Simplify $|5 - 7| - |3 - 8|$.

Q6) Solve the following inequality and graph the solution set on a number line:

$|3x - 1| < 11$.

Q7) Solve $|3x + 2| = |2x + 3|$.

Q8) Solve the following inequality:

$|x - 9| + 2 \leq 2$.

Q9) Solve $|3x| + 36 \geq 12$.

Q10) Solve the inequality $|4x - 8| > 12$, then graph the solution.

Chapter 7

Functions and Complex Numbers

Section 7.1 Introduction to Functions

Relation: A relation is a set of ordered pairs (x, y) of real numbers as in the following:

F = {(3, 2) (4, 1) (2, 4) (1, 3).

Domain and Range

The graph of a function can be used to determine the function's domain and range.

Domain: A set of inputs (found on the x-axis the collection of all x values in the graph).

Range: A set of outputs (found on the y-axis the collection of all y values in the graph).

In a relation, the set of all values of the independent variable (**x values**) is called the **Domain**.

In a relation, the set of all values of the dependent variable (**y value**s) is called the **Range**. If

F = {(3 , 5) (4 , 11) (2 , 27) (1 , 108)}, then the

Domain = {3, 4, 2, 1}   The domain is the set of all first components (x values).

Range = {5, 11, 27, 108}   The range is the set of all second components (y values).

Function: A function is a rule which maps a number to another unique number. In other words, a function is a relation in which, for each value of the first component, there is exactly one value of the second component. The following shape is a function expressed as a mapping:

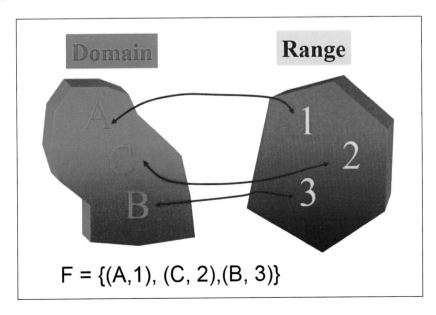

F = {(A,1), (C, 2),(B, 3)}

The following shape is not a function since "A" is mapped to two ranges "1" and "4":

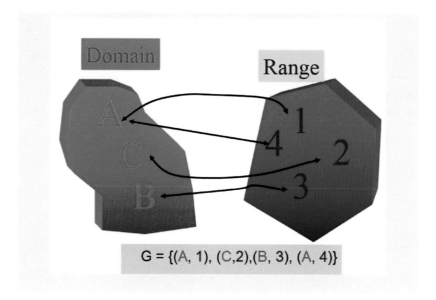

G = {(A, 1), (C,2),(B, 3), (A, 4)}

Example 1) Consider the relation {(3, 1), (5, 2), (3, 3), (7, 2)}. Find the domain and range of this relation. Is the relation a function?

Solution:

Domain = {3, 5, 7}

Range = {1, 2, 3}

Since the domain 3 goes to both 1 and 3, we know this relation is not a function.

Example 2) Consider the relation {(1, 5), (2, 6), (3, 7), (4, 8)}. Find the domain and range of this relation. Is the relation a function?

Solution:

Domain = {1, 2, 3, 4}

Range = {5, 6, 7, 8}

Since for each value of the first component there is exactly one value of the second component, this relation is a function.

Example 3) Determine whether each of the following relation is a function:

1) {(2, 3), (4, 8), (6, 8), (9, 10)}.

2) {(1, 2), (3, 3), (6, 8), (1, 10)}.

3) {(1, 5), (4, 5), (2, 5), (3, 5)}.

<u>Solution:</u>

    1) Yes – a function
    2) No – not a function since 1 is mapped to 2 and 10
    3) Yes – a function

Function Notation

When a function can be written as an equation, the symbol f(x) replaces y and it reads as "the value of f at x" Function notation also defines the value of x that is to be used to calculate the corresponding value of y. See the following:

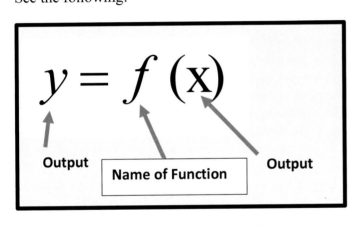

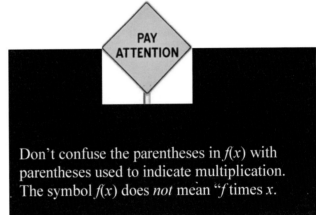

Don't confuse the parentheses in *f*(x) with parentheses used to indicate multiplication. The symbol *f*(x) does *not* mean "*f* times x."

> **Function Notation Notes**

1) When a function is defined by an equation, it is convenient to name the function.

2) Replacing y with *f*(x) is called writing a function in function notation.

3) The notation $y = f(x)$ provides a way of denoting the value of y (the dependent variable) that corresponds to some input number x (the independent variable).

4) An equation that is a function may be expressed using functional notation.

5) The notation *f*(x) (read "*f* of (x)") represents the variable y.

6) A function *f* of a variable x represented by *f*(x).

7) Functions can be evaluated at values and variables.

8) To evaluate a function, substitute the values for the domain for all occurrences of x.

9) (x, (*f*(x)) is an ordered pair of a function.

10) The inputs and outputs are called variables.

To evaluate f (2) in f(x) = x + 1, replace all x's with 2 and simplify:

f (2) = (2) + 1 = 3. This means that f (2) = 3.

Example 4) If $f(x) = 3x - 2$, find

  1) $f(4)$
  2) $f(5)$
  3) $f(0)$
  4) $f(-6)$

Solution:

$f(x) = 3x - 2$

$f(x) = 3 * (x) - 2$ notice that we can write 3x as 3 * (x)

$f(4) = 3 * (4) - 2 = 12 - 2 = 10$

$f(x) = 3 * (x) - 2$

$f(5) = 3 * (5) - 2 = 15 - 2 = 13$

$f(x) = 3 * (x) - 2$

$f(0) = 3 * (0) - 2 = 0 - 2 = -2$

$f(x) = 3 * (x) - 2$

$f(-6) = 3 * (-6) - 2 = -18 - 2 = -20.$

Example 5) If $g(x) = x^2 + 5$, then find g (-2), g (1), and g (3)

Solution:

$g(x) = (x)^2 + 5$

$g(-2) = (-2)^2 + 5 = 4 + 5 = 9$

$g(1) = (1)^2 + 5 = 1 + 5 = 6$

$g(3) = (3)^2 + 5 = 9 + 5 = 14.$

Q1) Consider the relation {(3, 1), (4, 2), (5, 0), (6, 7)}. Find the domain and range of this relation. Is the relation a function?

Q2) Consider the relation {(4, 1), (2, 2), (4, 3), (4, 2)}. Find the domain and range of this relation. Is the relation a function?

Q3) If $f(x) = 7.5x$, find $f(0)$.

Q4) The figure shows the cost of mailing a first-class letter, f(x), as a function of its weight, x, in ounces. Use the graph to answer the following questions.

    a) Find $f(3)$. What does this mean in terms of the variables in this situation?

    b) Find $f(4)$. What does this mean in terms of the variables in this situation?

    c) Find $f(5)$. What does this mean in terms of the variables in this situation?

    d) Find $f(6)$. What does this mean in terms of the variables in this situation?

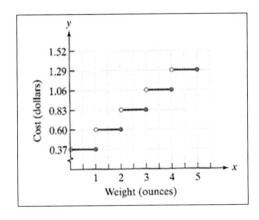

Q5) If $g(x) = g(x) = 2x^2 - 3x + 5$, find $g(1)$ and $g(-1)$.

Q6) If $g(x) = -6x + 9$

    a) $g(-4)$.

    b) $g(2)$.

    c) $g(3)$.

    d) $g(-20)$.

Section 7.2 Function Operations

Given two functions f and g, then for all values of x for which both $f(x)$ and $g(x)$ are defined, the functions
$f + g, f - g, f*g,$ and f/g are defined as follows:

$$(f + g)(x) = f(x) + g(x) \quad \text{Sum of two functions}$$

$$(f - g)(x) = f(x) - g(x) \quad \text{Difference of two functions}$$

$$(f * g)(x) = f(x) * g(x) \quad \text{Product of two functions}$$

$$\left(\frac{f}{g}\right)(x) = \frac{f(x)}{g(x)} \quad \text{Quotient of two functions}$$

The condition $g(x) \neq 0$ in the definition above of the quotient means that the domain of $(f/g)(x)$ is restricted to all values of x for which $g(x)$ is not 0. The condition does not mean that $g(x)$ is a function that is never zero.

You can perform operations on functions in much the same way that you perform operations on numbers or expressions. You can add, subtract, multiply, or divide functions by operating on their rules.

Key Points

➤ To add, subtract, multiply or divide functions just do it the same way you do any of these operations with numbers and expressions.

➤ The domain of the new function will have the restrictions of both functions that made it.

➤ Quotient has the additional tenet that the function we are dividing by cannot be zero.

Example 1) If $f(x) = 3x + 1$ and $g(x) = 5$, find $(f + g)(x)$.

Solution:

$(f + g)(x) = f(x) + g(x)$ sum of two functions.

$= 3x + 1 + 5$ substitute function rules (we replace $f(x)$ with $3x + 1$ and we replace $g(x)$ with 5).

$= 3x + 6.$

Example 2) If $f(x) = 7x + 4$ and $g(x) = 3x + 5$, find $(f + g)\ (x)$.

<u>Solution:</u>

$(f + g)\ (x) = f(x) + g(x)$ sum of two functions

        $= 7x + 4 + 3x + 5$ here we replace $f(x)$ with $7x + 4$ and we replace $g(x)$ with $3x+5$

        $= 10x + 9$ we combine the like terms: $7x + 3x = 10x$ and $4 + 5 = 9$.

Example 3) Given $f(x) = 8x^2 + 8x - 2$ and $g(x) = 7x^2 + 6x + 5$, find

1) $f(x + g)$.

2) $f(x - g)$.

<u>Solution:</u>

1) $(f + g)\ (x) = \quad f(x) \quad + \quad g(x)$

        $= 8x^2 + 8x - 2 \quad + \quad 7x^2 + 6x + 5$ here we combine $8x^2 + 7x^2 = 15x^2$,

                                        $8x + 6x = 14x$,

                                        and $-2 + 5 = +3$

        $= 15x^2 + 14x + 3$

2) $(f - g)\ (x) = \quad f(x) \quad - \quad g(x)$

        $= 8x^2 + 8x - 2 \ - (7x^2 + 6x + 5)$

        $= 8x^2 + 8x - 2 \ - 7x^2 - 6x - 5$ here we combine $8x^2 - 7x^2 = x^2$,

                                          $8x - 6x = 2x$,

                                        and $-2-5 = -7$

        $= x^2 + 2x - 7$.

Example 4) Given $f(x) = 4x + 1$ and $g(x) = 3x + 5$, find $(f - g)\ (2)$ and $(f + g)\ (2)$.

<u>Solution:</u>

$(f - g)\ (2) = f\ (2) - g\ (2)$

We need to find $f\ (2)$ and $g\ (2)$.

$f\ (x) = 4x + 1$

$f\ (2) = 4(2) + 1 = 8 + 1 = 9$

$g\ (x) = 3x + 5$

$g\ (2) = 3(2) + 5 = 6 + 5 = 11$

$(f - g)\ (2) = f\ (2) - g\ (2) = 9 - 11 = -2$

$(f + g)\ (2) = f\ (2) + g\ (2) = 9 + 11 = 20.$

Example 5) Given $f(x) = -2x - 1$ and $g(x) = \dfrac{2}{5}x + 5$, find $(f - g)\ (0)$ and $(f + g)\ (0)$.

<u>Solution:</u>

$(f - g)\ (0) = f\ (0) - g\ (0)$

We need to find $f\ (0)$ and $g\ (0)$.

$f\ (x) = -2x - 1$

$f\ (0) = -2(0) - 1 = 0 - 1 = -1$

$g\ (x) = \dfrac{2}{5}x + 5$

$g\ (0) = \dfrac{2}{5}(0) + 5 = 0 + 5 = 5$

$(f - g)\ (0) = f\ (0) - g\ (0) = -1 - 5 = -6$

$(f + g)\ (0) = f\ (0) + g\ (0) = -1 + 5 = +4.$

Example 6) Given $f(x) = 5$ and $g(x) = 6$, find $(f * g)(x)$.

Solution:

$(f * g)(x) = f(x) * g(x)$ Multiplication of two functions

$= 5 * 6 = 30.$

Example 7) Given $f(x) = x+2$ and $g(x) = x+3$, find $(f * g)(x)$.

Solution:

$(f * g)(x) = f(x) * g(x).$ Multiplication of two functions.

$= (x+2) * (x+3) = x^2 + 5x + 6.$ Here we use foil method.

Example 8) Given $f(x) = 18$ and $g(x) = 6$, find $(f / g)(x)$.

Solution:

$(f / g)(x) = f(x) / g(x)$ Quotient of two functions

$= 18 / 6 = 3.$

Example 9) Given $f(x) = x^2 + 5x + 6$ and $g(x) = x+2$, find $(f/g)(x)$.

Solution:

$$(f / g)(x) = \frac{f(x)}{g(x)} = \frac{x^2 + 5x + 6}{(x + 2)} = \frac{(x + 3)(x + 2)}{(x + 2)} = x + 3$$

Example 10) Given $f(x) = \sqrt{81}$ and $g(x) = |-3|$, find $(f/g)(x)$.

Solution:

$$(f / g)(x) = \frac{f(x)}{g(x)} = \frac{\sqrt{81}}{|-3|} = \frac{9}{3} = 3.$$

213

**H
O
M
E
W
O
R
K**

Q1) If $f(x) = x + 1$ and $g(x) = 5$, find $(f + g)(x)$.

Q2) If $f(x) = x + 1$ and $g(x) = 3x + 8$, find $(f + g)(0)$.

Q3) Given $h(t) = 2t - 4$ and $g(t) = t^2 + 5$, find $h(t) - g(t)$.

Q4) Given $h(t) = 2t + 14$ and $g(t) = t^2 + 15$, find $(h - g)(-1)$.

Q5) If $h(n) = -5$ and $g(n) = 8$, find $(h + g)(n)$.

Q6) If $f(x) = 2x - 1$ and $g(x) = 3x + 5$, find $(f * g)(x)$.

Q7) Given $f(x) = x - 4$ and $g(x) = x + 4$, find $(f * g)(5)$.

Q8) Given $f(x) = x$ and $g(x) = x + 4$, find $3*f(x)*g(x)$.

Q9) Given $f(x) = x^2 - 4x - 32$ and $g(x) = x + 4$, find $(f / g)(x)$.

Q10) Given $f(x) = x^2 + x - 6$ and $g(x) = \dfrac{x - 2}{5}$, find $(f / g)(-4)$.

Section 7.3 Composition of Functions and Inverse

Composition of Functions

If f and g are functions, then the composite function, or composition, of g and f is defined by:

$$(g \circ f)(x) = g(f(x)).$$

<u>Notes:</u>

1) Composition of functions is just combining two or more functions, but evaluating them in a certain order. It's almost like one is inside the other.
2) Composition of functions is the successive application of the functions in a specific order.
3) Given two functions f and g, the **composite function** is defined by $f(g(x))$ or $(f \circ g)(x)$ and is read "f of g of x."
4) The domain of $f \circ g$ is the set of elements x in the domain of g such that g(x) is in the domain of f. In other words, the range of function g must be in the domain of function f.

Let's say that Lana found two coupons for her favorite clothing store: One is a 25% discount, and another one is $15 off. The store allows her to use both in any order. She needs to figure out which way is the better deal if she bought $100 worth of clothes?

The following table describes what happens when she applies both options of the discounts in different orders if she was to buy $100 worth of clothes:

Order of Discounts	Result	Composition of Functions
Option 1) $15off first, followed by 25% off.	$g(x) = x - 15$ $g\ (100) = 100 - 15 = \$85$ $f(x) = 0.75 * x$ $f(85) = 0.75 * 85 =$ $\$63.75$	$(f \circ g)(x) = f(g(x))$ $\quad = f(x\text{-}15)$ $\quad = 0.75(x\text{-}15) = 0.75x\ \text{-}11.25$ $f(g\ (100)) = 0.75(100) - 11.25 =$ $\$63.75$
Option 2) 25% off first, followed by $15 off	$f(x) = 0.75 * x$ $f(100) = 0.75*100 = \$75$ $g(x) = x - 15$ $g\ (75) = 75 - 15 = \$60$	$(g \circ f)(x) = g(f(x))$ $\quad = g(0.75x) = 0.75x -$ 15 $g\ (f\ (100) = 0.75(100)\text{-}15 = \$60$

From the above table, it'd be better for Lana to take the 25% off first.

Example 1) Given

$$f(x) = x - 3 \text{ and}$$

$$g(x) = 2x^2 - 1,$$

Evaluate $(g \circ f)(x)$.

Solution:

$$(g \circ f)(x) = g(f(x))$$

$$= 2(x - 3)^2 - 1$$

$$= 2(x^2 - 6x + 9) - 1$$

$$= 2x^2 - 12x + 18 - 1$$

$$== 2x^2 - 12x + 17.$$

Example 2) If $f(\mathbf{x}) = \mathbf{x}+5$ and $g(\mathbf{x}) = \mathbf{8}$, find $(f \circ g)(x)$.

Solution:

$$(f \circ g)(x) = f(g(\mathbf{x}))$$

$$= f(\mathbf{8})$$

$$= \mathbf{8} + 5$$

$$= 13.$$

Example 3) If $f(\mathbf{x}) = \mathbf{x} + 5$ and $g(\mathbf{x}) = \mathbf{-4}$, find $(f \circ g)(x)$

Solution:

$$(f \circ g)(x) = f(g(\mathbf{x}))$$

$$= f(\mathbf{-4})$$

$$= \mathbf{-4} + 5$$

$$= 1.$$

Example 4) If $f(x) = x^2 + 3x$, $g(x) = 2x - 3$ find $(f \circ g)(-2)$.

Solution:

$$(f \circ g)(-2) = f(g(-2))$$

$$g(-2) = 2(-2) - 3 = -4 - 3 = -7$$

$$(f \circ g)(-2) = f(g(-2)) = f(-7)$$

$$= (-7)^2 + 3(-7)$$

$$= 49 - 21 = 28$$

Inverse of a Function

Definition:

Let f and g be two functions such that
$$f(g(x)) = x \quad \text{for every } x \text{ in the domain of } g$$

and $\quad g(f(x)) = x \quad$ for every x in the domain of f.

The function g is the inverse of the function f, and is denoted by f^{-1} (read "f-inverse"). Thus,

$$f(f^{-1}(x)) = x \text{ and } f^{-1}(f(x)) = x.$$

The domain of f is equal to the range of f^{-1} and vice versa.

Reminder:

A function is a relation in which no member of the domain is repeated.

Steps for finding the equation for the inverse of a function f:

Step# 1: Replace $f(x)$ with y. Remember that x is the domain, y is the range.

Step# 2: Interchange x and y (switch the x and y in the equation).

Step # 3: Solve for y. (Note: If y is not a function of x, then f does not have an inverse).

Step# 4: Replace y with $f^{-1}(x)$.

Example 5) Given $f(x) = 3x - 4$ find its inverse $(f^{-1}(x))$.

<u>Solution:</u>

$f(x) = x + 5$

1) Replace $f(x)$ with y

    $y = x + 5$

2) Switch x and y

    $x = y + 5$

3) Solve for y

    $x - 5 = y$

Note:

$f^{-1}(x) \neq f(x)$. It is very important not to confuse function notation with negative exponents.

4) Replace *y* with $f^{-1}(x)$

    $f^{-1}(x) = x - 5$.

Example 6) Find the inverse of the following function:

$f(x) = 3x - 4$.

<u>Solution:</u>

$f(x) = 3x - 4$

1) Replace $f(x)$ with y

    $y = 3x - 4$

2) Switch x and y

    $x = 3y - 4$

3) Solve for y

    $x + 4 = 3y$

    $y = \dfrac{x + 4}{3}$.

4) Replace *y* with $f^{-1}(x)$.

    $f^{-1}(x) = \dfrac{x + 4}{3}$.

Q1) Given $f(x) = x + 5$ and $g(x) = 9$ find $(f \circ g)(x)$.

Q2) If $f(x) = 3x - 4$ and $g(x) = x^2 + 6$, evaluate each of the following including their domains.

    (a) $f \circ g\ (5)$ **(b)** $g \circ f\ (5)$

Q3) Given $f(x) = 2x - 1$ and $g(x) = \dfrac{4}{x-1}$, find $(f \circ g)(2)$.

Q4) Evaluate each of the following indicated function without finding an equation for the function given:

    $f(x) = x + 1$ $g(x) = -2x$ $k(x) = x^2 - 3$

    (a) $f(g(0))$ **(b)** $(k \circ g)(4)$ **(c)** $f(f(10))$ **(d)** $f(g[k(2)])$ **(e)** $f \circ g\ (0)$

Q5) Given the following function:

    $f(x) = 3x^2 + 2$, find its inverse $(f^{-1}(x))$.

Q6) Find the inverse of $f(x) = x^5$. Is $f^{-1}(x)$ a function?

Q7) Find an equation for $f^{-1}(x)$ given $f(x) = (x-3)^3$.

    a) $f^{-1}(x) = \sqrt[3]{x} + 3$

    b) $f^{-1}(x) = \sqrt[3]{x+3}$

    c) $f^{-1}(x) = \sqrt[3]{x} - 3$

    d) $f^{-1}(x) = \sqrt[3]{x-3}$

Q8) If $k(x) = 2x - 5$, find the inverse.

Q9) If $f(x) = (x-2)^3$, find the inverse.

Q10) If $h(x) = \dfrac{4x-3}{2x+1}$, find $h^{-1}(x)$.

The set of all numbers in the form **a + b*i*** with real numbers "*a*" and "*b*", and "*i*", the imaginary unit, is called the set of **complex numbers** (see the shape below). The real number *a* is called the **real part**. The real number *b* is called the **imaginary part** of the complex number **a + b*i*** (standard form).

If **b** $\neq$ 0, then the complex number is called an imaginary number.

Thus, the real part of 4 - 3*i* is 4 and the imaginary part is -3.

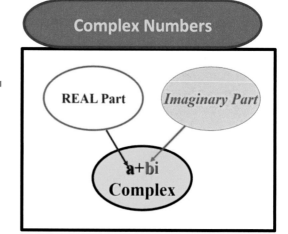

Equality of Complex Numbers

If $a + bi = c + di$, then $a = c$ and $b = d$.

Example 1) Solve the following equation for x and y:

$x + 5i = 7 + yi$

Solution:

$x + 5i = 7 + yi$

$x = 7$

$y = 5.$

Example 2) Solve the following equation for x and y:

$x - 3i = -4 + yi$

Solution:

$x - 3i = -4 + yi$

$x = -4$

$y = -3.$

Notes:

1) All real numbers are complex for example, $8 = 8 + 0i$.
2) All imaginary numbers are complex for example, $9i = 0 + 9i$.

The Imaginary Unit i

The imaginary unit i is defined as:

$$i = \sqrt{-1} \quad \text{where} \quad i^2 = -1.$$

The Square Root Property of a Negative Number

If x is any positive real number, then

$$\sqrt{-x} = \sqrt{(x)*(-1)} = \sqrt{x} * \sqrt{-1} = \sqrt{x}\ i.$$

Thus, $\sqrt{-5} = \sqrt{(5)*(-1)} = \sqrt{5} * \sqrt{-1} = \sqrt{5}\ i.$

Example 3) Write the following as a multiple of i:

$$\sqrt{-64}.$$

Solution:

$$\sqrt{-64} = \sqrt{(64)*(-1)} = \sqrt{64} * \sqrt{-1} = 8*i = 8i.$$

> **Reminder:**
> When working with negative radicands, use the definition…
> $$\sqrt{-x} = \sqrt{x}\ i$$
> before using any of the other rules for radicands.

Example 4) Express the following number in terms of i:

$$\sqrt{-25}.$$

Solution:

$$\sqrt{-25} = \sqrt{(25)*(-1)} = \sqrt{25} * \sqrt{-1} = 5*i = 5i.$$

Example 5) Write the following as a multiple of i:

$$\sqrt{-21}.$$

Solution:

$$\sqrt{-21} = \sqrt{(21)*(-1)} = \sqrt{21} * \sqrt{-1} = \sqrt{21}*i.$$

If you square a radical, you will get a radicand as in the following examples:

$$\left(\sqrt{5}\right)^2 = 5 \text{ and}$$

$$\left(\sqrt{7}\right)^2 = 7.$$

Powers of i

i is a symbol for a specific number. It is not a variable. When you have i to the power of any positive whole number greater or equal to 4, always divide the exponents by 4. If you get a remainder of **zero** (no remainder), the answer is **1** for example, $i^{12} = 1$ because when you divide 12 by 4 there will be no remainder. If you get a remainder of **1**, the answer is i. If you get a remainder of **2**, the answer is **-1**. If you get a remainder of **3**, the answer is $-i$. See the table below:

$i^1 = i$	$i^5 = i$	$i^9 = i$
$i^2 = -1$	$i^6 = -1$	$i^{10} = -1$
$i^3 = -i$	$i^7 = -i$	$i^{11} = -i$
$i^4 = 1$	$i^8 = 1$	$i^{12} = 1$

Example 6) Find the followings:

a) i^{200} b) i^{201} c) i^{2006} d) i^{23}

Solution:

a) $i^{200} = 1$ b) $i^{201} = i$ c) $i^{2006} = -1$ d) $i^{23} = -i$

Adding & Subtracting Complex Numbers

You can add complex numbers by:

    1) Adding their real parts
    2) Adding their imaginary parts
    3) Expressing the sum as a complex number

$$(a + bi) + (c + di) = (a + c) + (b + d)\, i$$

You can subtract complex numbers by:

    1) Subtracting their real parts
    2) Subtracting their imaginary parts
    3) Expressing the difference as a complex number

$$(a + bi) - (c + di) = (a - c) + (b - d)\, i$$

Example 7) Add the following:

$(1 + 2i) + (4 + 7i).$

Solution:

$$(1 + 2i) + (4 + 7i) = (1 + 4) + (2 + 7)\, i$$
$$= 5 \quad + \quad 9i.$$

Example 8) Subtract the following:

$(9 + 8i) - (5 + 6i).$

Solution:

$$(9 + 8i) - (5 + 6i) = (9 - 5) + (8 - 6)\, i$$
$$= 4 \quad + \quad 2i.$$

Example 9) Subtract the following:

$(8 + 6i) - (-2 - 11i).$

Solution:

$$(8 + 6i) - (-2 - 11i) = (8 - -2) + (6 - -11)\, i$$
$$= 8 + 2 \quad + (6 + 11)\, i$$
$$= 10 \quad + \quad 17i.$$

Multiplication of Complex Numbers

To multiply complex numbers, you use the same procedure as multiplying monomials or polynomials.

To multiply imaginary number (or numbers) by a real number, it is important to express the imaginary numbers in terms of i first.

Example 10) Multiply the followings:

        a) $8i * 7$

        b) $3(4i - 5)$

        c) $4i(-6+i)$

Solution:

        a) $8i * 7 = 56i$

        b) $3(4i - 5) = 12i - 15$ Distributive Property.

        c) $4i(-6+i) = -24i + 4i^2$ Distributive Property.

                $= -24i + 4(-1)$ Use $i^2 = -1$.

                $= -24i - 4$ Simplify.

                $= -4 - 24i$ Write in standard form.

Example 11) Multiply $(4 + 3i)(7 + 2i)$.

Solution:

$(4 + 3i)(7 + 2i) = 28 + 8i + 21i + 6i^2$ Multiply using FOIL Method.

              $= 28 + 29i + 6(-1)$ Simplify and use $i^2 = -1$.

              $= 28 + 29i - 6$ Simplify.

              $= 22 + 29i$ Write in standard form.

Example 12) Multiply $(9 - 2i)(-4 + 7i)$

Solution:

$(9 - 2i)(-4 + 7i) = -36 + 63i + 8i - 14i^2$ Multiply using FOIL Method.

              $= -36 + 71i - 14(-1)$ Simplify and use $i^2 = -1$.

              $= -36 + 71i + 14$ Simplify.

              $= -22 + 71i$ Write in standard form.

Conjugate of a Complex Number

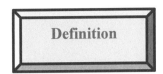

Definition

The conjugate of the complex number z = a + bi is denoted by and is given by $\overline{z}$ = a − bi.

From this definition, we can see that the conjugate of a complex number which is found by changing the sign of the imaginary part of the number.

> Conjugate means you change the sign between the two terms.

Example 13) Find the conjugate of each of the following complex numbers:

    a) z = 3 −7i

    b) z = -4 +15i

    c) z = -8i

    d) z = 9

Solution:

Complex Number	Conjugate
a) z = 3 −7i	$\overline{z}$ = 3 + 7i
b) z = -4 +15i	$\overline{z}$ = -4 − 15i
c) z = −8i	$\overline{z}$ = + 8i
d) z = 9	$\overline{z}$ = 9 Because 9 = 9 + 0i so the conjugate is 9 − 0i = 9

Properties of Complex Conjugates

For a complex number z = a + bi, the following properties are true:

1) $z\overline{z} = a^2 + b^2$

2) $z\overline{z} \geq 0$

3) $z\overline{z} = 0$ if and only if z = 0

4) $\overline{(\overline{z})} = z$

To prove the first property is right, we let z = a + bi then $\bar{z}$ = a − bi.

$$z\bar{z} = (a+bi)(a-bi) = a^2 - abi + abi - b^2i^2 = a^2 - b^2(-1) = a^2 + b^2$$

The second and third properties follow directly from the first. Finally, the fourth property follows the definition of complex conjugate. That is: $\overline{(\bar{z})} = \overline{(a+bi)} = \overline{a - bi} = a + bi = z$

Example 14) Find the product of z = 3 −7i and its complex conjugate.

Solution:

z = 3 −7i then $\bar{z}$ = 3 + 7i.

$$z\bar{z} = (3 - 7i)(3 + 7i) = 3^2 + 7^2 \quad \text{Property\# 1 } (z\bar{z} = a^2 + b^2)$$

$$= 9 + 49 = 58.$$

The Modulus of a Complex Number

The modulus of the complex number w = a + bi is denoted $|w|$ and is given by:

$$|w| = \sqrt{a^2 + b^2}.$$

Note: The modulus of a complex number is also called the absolute value of the number. Indeed, when w = 3+4i is a real number, we have $|w| = \sqrt{3^2 + 4^2} = \sqrt{9 + 16} = \sqrt{25} = 5$

Example 15) For w = 2 +i and z = 2 − i, determine the following:

a) $|w|$ b) $|z|$ c) $2 * |w| * |z|$ d) $|wz|$

Solution:

a) $|w| = \sqrt{2^2 + 1^2} = \sqrt{5}$

b) $|z| = \sqrt{2^2 + (-1)^2} = \sqrt{5}$

c) $2 * |w| * |z| = 2 * \sqrt{5} * \sqrt{5} = 2\sqrt{25} = 2 * 5 = 10$

d) Since wz = (2+i) (2−i) = $2^2 + 1^2$ = 4+1 = 5 = 5 + 0i, we have $|wz| = \sqrt{5^2 + 0^2} = \sqrt{25} = 5$.

Division of Complex Numbers

One of the most significant uses of the conjugate of a complex number is in performing division in the complex number system. To define division of complex numbers, let us consider $w = a + bi$ and $z = c + di$, where c and d are not both equal to 0.

If the quotient $\dfrac{w}{z} = x + yi$, it will lead to $w = z(x + yi) = (c + di) * (x + yi) = (cx - dy) + (dx + cy)\,i.$

But since $w = a + bi$, we can form the following linear system:

$\quad a = cx - dy$

$\quad b = dx + cy$

Solving this system of linear equations for x and y we get:

$$x = \frac{ac + bd}{z\bar{z}} \quad \text{and} \quad y = \frac{bc - ad}{z\bar{z}}$$

Now since, $w\bar{z} = (a + bi)(c - di) = (ac + bd) + (bc - ad)\,i$, we get the following:

The division of the complex numbers $w = a + bi$ and $z = c + di$ is defined to be

$$\frac{w}{z} = \frac{a + bi}{c + di} = \frac{a + bi}{c + di} * \frac{c - di}{c - di} = \frac{ac + bd}{c^2 + d^2} + \frac{bc - ad}{c^2 + d^2}\,i = \frac{1}{|z|^2} * (w\bar{z}), \text{ where } c^2 + d^2 \neq 0$$

Note:
When $c^2 + d^2 = 0$ then $c = d = 0$, and therefore $z = 0$. The quotient of two complex numbers can be found by multiplying the numerator and the denominator by the conjugate of the denominator, as follows:

$$\frac{a + bi}{c + di} = \frac{a + bi}{c + di} * \left(\frac{c - di}{c - di}\right) = \frac{(a + bi)(c - di)}{(c + di)(c - di)} = \frac{(ac + bd)(bc - ad)}{c^2 + d^2} = \frac{ac + bd}{c^2 + d^2} + \frac{bc - ad}{c^2 + d^2}\,i.$$

Rationalizing Complex Denominator

To rationalize a complex denominator (re-write a fraction so the bottom is a rational number), do the following steps:

1) Multiply the numerator and denominator by the complex conjugate of the denominator.

2) Change $i^2 = -1$.

3) Simplify and combine like terms when possible.

4) Change it to a standard form (a + bi).

Example 16) Rationalize the denominator of the following complex fraction:

$$\frac{7+5i}{1-4i}.$$

<u>Solution:</u>

$\frac{7+5i}{1-4i} = \frac{7+5i}{1-4i} * \frac{1+4i}{1+4i}$ Multiply numerator and denominator by 1 + 4i (which is the complex conjugate of 1 – 4i).

$= \frac{7+28i+5i+20i^2}{1+4i-4i-16i^2}$ Multiply using Foil Method.

$= \frac{7+33i+20(-1)}{1-16(-1)}$ Simplify and use $i^2 = -1$.

$= \frac{-13+33i}{17}$ Simplify.

$= \frac{-13}{17} + \frac{33i}{17}$ Write in standard form.

> **Note:**
>
> A complex number written in standard form is a number a + bi where a and b are real numbers.

Example 17) Write the following expression as a complex number in standard form.

$$2i\,(3 - 7i).$$

Solution:

$2i\,(3 - 7i) = 6i - 14\,i^2$ Distributive property.

$\quad\quad\quad = 6i - 14(-1)$ Use $i^2 = -1$.

$\quad\quad\quad = 6i + 14$ Simplify.

$\quad\quad\quad = 14 + 6i$ Write in standard form.

Example 18) Simplify the following expression. Write the answer in standard form.

$$\frac{1}{i}.$$

Solution:

$\dfrac{1}{i} = \dfrac{1}{i} * \dfrac{-i}{-i} = \dfrac{-i}{-i^2}$ Multiply numerator and denominator by $-i$ (which is the complex conjugate of i).

$\quad\quad = \dfrac{-i}{-(-1)}$ Simplify and use $i^2 = -1$.

$\quad\quad = \dfrac{-i}{1} = -i$ Simplify.

Example 19) Simplify the following expression. Write the answer in standard form.

$$\frac{3}{1+i}.$$

Solution:

$\dfrac{3}{1+i} = \dfrac{3}{1+i} * \dfrac{1-i}{1-i}$ Multiply numerator and denominator by $1 - i$ (which is the complex conjugate of $1+i$).

$\quad\quad = \dfrac{3 - 3i}{1 - i^2}$ Multiply.

$\quad\quad = \dfrac{3 - 3i}{1 - (-1)} = \dfrac{3 - 3i}{2} = \dfrac{3}{2} - \dfrac{3}{2}i$ Simplify and use $i^2 = -1$.

Example 20) Write the following quotient in standard form $a + bi$:

$$\frac{3+2i}{5-i}.$$

<u>Solution:</u>

$\frac{3+2i}{5-i} = \frac{3+2i}{5-i} * \frac{5+i}{5+i}$ Multiply numerator and denominator by $1 + 4i$ (which is the complex conjugate of $1 - 4i$).

$= \frac{15+3i+10i+2i^2}{25+5i-5i-i^2}$ Multiply using Foil Method.

$= \frac{15+13i+2(-1)}{25-(-1)}$ Simplify and use $i^2 = -1$.

$= \frac{13+13i}{26}$ Simplify.

$= \frac{13}{26} + \frac{13i}{26}$ Write in standard form.

$= \frac{1}{2} + \frac{1}{2}i$ Simplify.

Exercise Set 7.4

Q1) Solve the following equation for x and y:

$x - 22i = -1 + yi.$

Q2) Express the following number in terms of i:

$\sqrt{-44}.$

Q3) $i^{24} = ?$

Q4) Add the following complex numbers: $(1 - 2i) + (8 + 7i)$.

H
O
M
E
W
O
R
K

230

Q5) Subtract the following complex number:

$(9 - 2i) - (8 - 7i).$

Q6) Multiply the followings:

      a) $3i * 9$

      b) $-5(-2i - 11)$

      c) $2i(-6 - i)$

      d) $(3 - 8i)(9 + 6i)$

      e) $(9 - i)(9 + i)$

Q7) Find the conjugate of the following complex number:

$8 - 24i.$

Q8) For $w = 5 + 9i$ and $z = 4 - 7i$, determine the following:

    a) $|w|$ **b)** $|z|$ **c)** $2*|w|*$ $|z|$ **d)** $|wz|$ **e)** $|z| \div |w|$

Q9) Find the product of $z = 1 - 6i$ and its complex conjugate.

Q10) Rationalize the denominator of the following complex fraction:

$$\frac{8 - 6i}{7 - 3i}.$$

Q11) Write the following expression as a complex number in standard form $a + bi$.

$5i(-8 - 7i).$

Q12) Write the following quotient in standard form $a + bi$.

$$\frac{5}{i}.$$

Q13) Write $\dfrac{-8 + \sqrt{-128}}{4}$ in standard form $a + bi$.

Q14) Simplify the following:

$$\frac{4+2i}{3+i} + \frac{6-i}{4}.$$

Q15) Solve $x^2 - 2x = -2$ (leave your answer in standard form a +bi).

Q16) Evaluate each of the following expressions and write your answer in the standard form a +bi:

a) $(4 - \frac{1}{2}i) - (9 + \frac{5}{2}i)$ **b)** $(1 - 2i)(7 - 6i)$ **c)** $\frac{3+2i}{1-4i}$ **d)** i^{43} **e)** $\sqrt{-121}$

Q17) Write the following expression as a complex number in standard form:

$i(20 - i).$

Q18) Simplify the following expression:

a) $(4 - 4i) \div (3 - 4i^3)$

b) $(10 + 6i) \div (2 - i)$

Q19) $i^{200} = ?$

a) 1

b) i

c) –i

d) –1

Q20) Find the values of the real numbers "h" and "k" in each of the following:

a) $\frac{h}{1+i} + \frac{k}{1-2i} = 1$

b) $\frac{h}{2-i} + \frac{ki}{i+3} = \frac{2}{1+i}.$

Chapter 1

Section 1.1	Section 1.2	Section 1.3	Section 1.4
Q1) a) -9 b) -17 c) +18 d) +26	**Q1)** $\sqrt{5}$	**Q1)** Distributive property	**Q1)** 28/100
Q3) a) +20 b) +72 c) +12 d) +8	**Q3)** 0	**Q3)** Identity Property of Addition	**Q3)** 60
Q5) a) +28 b) +72 c) +48 d) +56	**Q5)** 1	**Q5)** Inverse Property of Multiplication	**Q5)** 7/12
Q7) 22	**Q7)** 0	**Q7)** Commutative and associative property of addition	**Q7)** 1
Q9) −35	**Q9)** 0	**Q9)** Commutative Property of Multiplication	**Q9)** 21/8
	Q11) The distance between -15 and -20 = $\|(-15)-(-20)\| =$ $\|-15+20\|=\|+5\|=5$		
	Q13) A U B = {0,1,2,3,4}		
	Q15) X ∩ Y = {5, 6, 7}		

Chapter 2

Section 2.1	Section 2.2	Section 2.3	Section 2.4	Section 2.5
Q1) 6	**Q1)** x	**Q1)** 5	**Q1)** 7.9	**Q1)** $x \geq 2$
Q3) $x - 4 = 8$	**Q3)** -19x-14	**Q3)** m = 2 b = -7	**Q3)** 162	**Q3)** $y < -6$
Q5) 88	**Q5)** $-2y^2 + 3y$	**Q5)** -5	**Q5)** 13	**Q5)** D
Q7) 5C = D		**Q7)** $y = \frac{1}{3}x + 2$	**Q7)** 15 hours	**Q7)** $x \leq 0$
Q9) T - 5		**Q9)** $x = \frac{1}{15a}$	**Q9)** 38	
		Q11) y= -2x – 1		
		Q13) y = 3x + 2		

Chapter 3

Section 3.1	Section 3.2	Section 3.3	Section 3.4

Q1)
  a. 1
  b. 10
  c. 100
  d. 1000
	Q1) $\frac{1}{x-2}$	**Q1)** $\frac{3x}{y^2} * \frac{4x}{y} = \frac{12x^2}{y^3}$	**Q1)** 1

Q3) $(x^2)^5 = x^{10}$	**Q3)** 5	**Q3)** x = {4, -7}	**Q3)** $\frac{C}{B^2 A^2}$
Q5) $\frac{x^{18}}{x^{15}} = x^3$	**Q5)** $(x-3)(x-2)^2$	**Q5)** $\frac{2}{3}$	
Q7) $\frac{1}{x}$	**Q7)** $\frac{12x + 7}{10x^2}$	**Q7)** $12x^3$	

Chapter 4

Section 4.1	Section 4.2	Section 4.3	Section 4.4
Q1) $6y + x + 7a$	**Q1)** $m^2 + m - 12$	**Q1)** 5	**Q1)** $x = 6$
Q3) $7x + 9y$	**Q3)** $r^2 - 13r + 40$	**Q3)** 4	**Q3)** $x = \dfrac{1}{2}$ $\quad or \quad x = \dfrac{-4}{3}$
Q5) $3x^2 + 8y^2 - 12x$	**Q5)** $4m^2 - 4mp + p^2$	**Q5)** $2^2 * 5^2$	**Q5)** $x = 3$ o r $x = -5$
Q7) Trinomial of degree 2	**Q7)** $8x^2 + 10x + 3$	**Q7)** x^2	**Q7)** $x = -7$ or $x = 9$
	Q9) $x^2 - 9$	**Q9)** $11x$	**Q9)** Length = 30 feet Width = 20 feet
			Q11) C

Chapter 5

Section 5.1	Section 5.2	Section 5.3	Section 5.4
Q1) $4\sqrt{5}$	**Q1)** $8\sqrt{15}$	**Q1)** $5\sqrt{2}$	**Q1)** $\left(4, \dfrac{-11}{2}\right)$
Q3) $7\sqrt{x}$	**Q3)** $\sqrt[5]{A^3 B^3}$	**Q3)** $4x^2 y\sqrt{3y}$	**Q3)** 5
Q5) 6	**Q5)** x^5	**Q5)** 13	**Q5)** $(1, 3)$
Q7) $-3\sqrt[3]{a}$	**Q7)** $\dfrac{5 + \sqrt{3}}{22}$	**Q7)** $\dfrac{5x\sqrt{2x}}{9y^4}$	**Q7)** 5.8
Q9) $2\sqrt[7]{x^2 y^5}$	**Q9)** $x = \pm\dfrac{9}{2}$	**Q9)** 81	
	Q11) $x = \pm\dfrac{4}{7}$		

Chapter 6

Section 6.1	Section 6.2	Section 6.3	Section 6.4	Section 6.5
Q1) 4 hrs. and 24 minutes	**Q1)** $1.3497 * 10^3$	**Q1)** $4.5 * 10^7$	**Q1)** (0, -2)	**Q1)** 7.5 hrs.
Q3) 16	**Q3)** $7.05 * 10^{-3}$	**Q3)** $8 * 10^7$	**Q3)** (5, 4)	**Q3)** 100 km/hr.
Q5) 37	**Q5)** $6.0 * 10^7$	**Q5)** $6.72 * 10^2$	**Q5)** 18 and 7	**Q5)** 36 mph
Q7) 4	**Q7)** $32.74 * 10^2$	**Q7)** $8.64 * 10^{10}$	**Q7)** 33 and 30	**Q7)** 9.25 hrs.
Q9) $x = \dfrac{5}{8}$	**Q9)** 1	**Q9)** $4 * 10^3$	**Q9)** 20 and 61	**Q9)** a. 45 words per minutes. b. 2700 words per hour.
			Q11) a	

Section 6.6	Section 6.7	Section 6.8
Q1) $1505	**Q1)** $x = \{\pm \sqrt{7}\}$	**Q1)** x = 4 and x = -3
Q3) $160	**Q3)** 11 feet	**Q3)** a = -1 and a = -9
Q5) $914.25	**Q5)** 48 feet	**Q5)** -3
Q7) 7.5%	**Q7)** $x = \{\pm 9\sqrt{2}\}$	**Q7)** x =1 and x = -1
Q9) 2.5	**Q9)** $x = \{\pm 5\}$	**Q9)** x ≥ -8 and x ≤ 8

Chapter 7

Section 7.1	Section 7.2	Section 7.3	Section 7.4	Section 7.4

Q1) Domain = {3, 4, 5, 6}
      Range = {1, 2, 0, 7}
Yes, it is a function.
	Q1) x + 6	**Q1)** 14	**Q1)** x = −1
			      y = −22
				Q11) 35 −40i

Q3) 0	**Q3)** $-t^2 + 2t - 9$	**Q3)** 7	**Q3)** $i = 1$	**Q13)** $-2 + 2\sqrt{2}\,i$
Q5) g(1) = 4 g(-1) = 10	**Q5)** 3	**Q5)** $f^{-1}(x) = \sqrt{\dfrac{x-2}{3}}$	**Q5)** 1 + 5i	**Q15)** $x = 1 \mp i$
	Q7) 9	**Q7)** a	**Q7)** 8+ 24i	**Q17)** 1 +20i
	Q9) x - 8	**Q9)** $f^{-1}(x) = \sqrt[3]{x} + 2$	**Q9)** 37	**Q19)** 1

- Akrayee, Dilshad (2007)

   Retrieved from the World Wide Web

   http://www.highlands.edu/site/faculty-dilshad-akrayee

- Jerry Howett (2001). Contemporary's GED Mathematics.

- Gilbert Strang (2009). Linear Algebra: A Modern Introduction (3rd Edition).

- Robert Mitchell, Contemporary's Number Power 3: Algebra A Real World Approach to Math (1st Edition).

- Kathleen Almy and Heather Foes (2013). Math Lit Plus MyMathLab.

- D. Franklin Wright (2012). Introductory & Intermediate Algebra.

- Charles P. McKeague (2004). Elementary and Intermediate Algebra.

- Stanley I. Grossman (1981). Calculus (2nd Edition).

- Sheldon Axler (2014), Linear Algebra Done Right (3rd Edition).

- Marvin Bittinger and David Ellenbogen (2005). Elementary Algebra. Concepts and Applications (7th Edition).

- Peter Selby & Steve Slavin. Practical Algebra: A Self-Teaching Guide (2nd Edition).

- David C. Lay (2011). Linear Algebra and Its Applications (4th Edition).

- Retrieved from the World Wide Web: http://bitsandpieces.us/2016/02/10/my-first-computer/

Made in the USA
Lexington, KY
21 August 2017